CCEA

GCSE
SCIENCE
SINGLE AWARD
SECOND EDITION

Helen Dowds
Alyn G McFarland
James Napier
Roy White

HODDER
EDUCATION
AN HACHETTE UK COMPANY

Photo credits

p.1 © D. Kucharski K. Kucharska/ Shutterstock; **p.4** © D. Kucharski K. Kucharska/ Shutterstock; **p.5** © Simon Fraser / Department of Haematology, RVI, Newcastle / Science Photo Library; **p.8** © Science Photo Library; **p.12** both © Science Photo Library; **p.16** © L. Willatt, East Anglian Regional Genetics Service / Science Photo Library; **p.17** © Girand / Science Photo Library; **p.18** l © L. Willatt, East Anglian Regional Genetics Service / Science Photo Library, r © CNRI / Science Photo Library; **p.22** © Saturn Stills / Science Photo Library; **p.24** © Volker Steger / Science Photo Library; **p.25** © MRAORAOR / Shutterstock; **p.27** © James Napier; **p.31** t © Science Photo Library, b © Sue Ford / Science Photo Library; **p.32** © James Napier; **p.35** © Adam Hart-Davis / Science Photo Library; **p.37** © Edelmann / Science Photo Library; **p.39** l © Adam Hart-Davis / Science Photo Library, r © TEK Image / Science Photo Library; **p.43** © Gecko1968 / Shutterstock; **p.44** © Ozgur Coskun / Shutterstock; **p.45** © Gecko1968 / Shutterstock; **p.46** © Bob Thomas / Popperfoto / Getty Images; **p.47** Chris Hellier / Science Photo Library; **p.48** © Interfoto / Alamy Stock Photo; **p.49** © holbox / Shutterstock; **p.51** © Tek Image / Science Photo Library; **p.52** © Kent Wood / Science Photo Library; **p.55** © John Durham / Science Photo Library; **p.56** t © St Mary's Hospital Medical School / Science Photo Library, bl © Science Photo Library, br © National Library of Medicine / Science Photo Library; **p.57** © Tek Image / Science Photo Library; **p.58** © Matt Cardy / Getty Images; **p.61** © JPL Designs / Shutterstock; **p.63** © JPL Designs / Shutterstock; **p.67** tl © Piotr Krzeslak / Shutterstock, tr © Michael G. Mill / Shutterstock, b © James Osmond Photography / Alamy Stock Photo; **p.68** t © TalyaPhoto/Shutterstock, b © Peter Gudella / Shutterstock; **p.73** © Bjoern Wylezich / Shutterstock; **p.74** © Maxx-Studio / Shutterstock; **p.75** © Mikkel Juul Jensen / Science Photo Library; **p.76** t © Helene Rogers / Alamy Stock Photos, b © Bjoern Wylezich / Shutterstock; **p.77** © Martyn F. Chillmaid / Science Photo Library; **p.78** © GIPhotoStock / Science Photo Library; **p.80** © Gregory Davies / Science Photo Library; **p.91** t © Claude Nuridsany and Marie Perennou / Science Photo Library, b © Martyn F. Chillmaid / Science Photo Library; **p.93** © Claude Nuridsany and Marie Perennou / Science Photo Library; **p.97** l © Trevor Clifford Photography / Science Photo Library, r © Andrew Lambert Photography / Science Photo Library; **p.100** © Colin Underhill / Alamy Stock Photo; **p.103** © Sputnik / Science Photo Library; p.116 © Tatiana Shepeleva / Shutterstock; **p.118** © aimy27feb / 123RF; **p.119** © Tatiana Shepeleva / Shutterstock; **p.122** © Science Photo Library / Alamy Stock Photo; **p.123** © arfo / Shutterstock; **p.124** t © Couperfield / Shutterstock; b Dung Vo Trung / Eurelios / Look at Sciences / Science Photo Library; **p.125** t © Chutima Chaochaiya / 123RF, b © Denys Prykhodov / Shutterstock; **p.127** GIPhotoStock / Science Photo Library; **p.128** all © Martyn F. Chillmaid / Science Photo Library; **p.130** © GIPhotoStock / Science Photo Library; **p.137** © Tracy Carnahan / Shutterstock; **p.143** © Dmitri Ma / Shutterstock; **p.146** © studioDG / Fotolia; p.153 t © Muellek Josef / Shutterstock, bl © Dmitri Ma / Shutterstock, br © Dirk Wiersma / Science Photo Library; **p.155** © PhotosIndia.com LLC / 123RF; **p.168** © Alakin Maksim Valerevich / Shutterstock; **p.172** © imagedb.com / Shutterstock; **p.173** © imagedb.com / Shutterstock; **p.179** © Steven Heap / 123RF; **p.187** © Chris Howes / Wild Places Photography / Alamy Stock Photo; **p.190** tl © photka / 123RF, tr © StockPhotosArt / Shutterstock, cl © Vladyslav Bashutskyy / 123RF, cr © Pavlo Herhelizhiu / 123RF, bl © Yuri Bizgajmer / 123RF, br © belchonock / 123RF; **p.191** t © Chris Howes / Wild Places Photography / Alamy Stock Photo, b © Anne Wanjie; **p.193** © Diyana Dimitrova / 123RF; **p.201** © Darren Brode / Fotolia; **p.206** © Darren Brode / Fotolia; **p.210** © Fernando Camino / Cover / Getty Images; **p.214** © Darryl Sleath / Shutterstock; **p.216** © Tony Camacho / Science Photo Library; **p.220** l © Darryl Sleath / Shutterstock, r © Adam Hart-Davis / Science Photo Library; **p.221** t © Tomas Skopal / Shutterstock, b © Paolo77/123rf; **p.228** © Thomas Fredberg / Science Photo Library; **p.230** © wellphoto / Shutterstock; **p.233** © GustoImages / Science Photo Library; **p.234** t © Thomas Fredberg, b © Cordelia Molloy / Science Photo Library; **p.237** t © World History Archive / Alamy Stock Photo, b © Friedrich Saurer / Science Photo Library; **p.238** t © Frank Zullo / Science Photo Library, b © Action Sports Photography / Shutterstock; **p.239** ©World History Archive / Alamy Stock Photo; **p.244** © NASA Images / Shutterstock.

Acknowledgements

Although every effort has been made to ensure that website addresses are correct at time of going to press, Hodder Education cannot be held responsible for the content of any website mentioned in this book. It is sometimes possible to find a relocated web page by typing in the address of the home page for a website in the URL window of your browser.

Hachette UK's policy is to use papers that are natural, renewable and recyclable products and made from wood grown in well-managed forests and other controlled sources. The logging and manufacturing processes are expected to conform to the environmental regulations of the country of origin. Orders: please contact Hachette UK Distribution, Hely Hutchinson Centre, Milton Road, Didcot, Oxfordshire, OX11 7HH. Telephone: +44 (0)1235 827827. Email education@hachette.co.uk. Lines are open from 9 a.m. to 5 p.m., Monday to Friday. You can also order through our website: www.hoddereducation.co.uk

© Helen Dowds, Alyn G McFarland, James Napier, Roy White

First edition published 2013
Second edition published 2017 by
Hodder Education,
An Hachette UK Company
Carmelite House
50 Victoria Embankment
London EC4Y 0DZ
Impression number 10 9 8
Year 2023

Cover photo © Seaphotoart / Alamy Stock Photo
Illustrations by Elektra Media Ltd
Typeset by Elektra Media Ltd
Printed by CPI Group (UK) Ltd, Croydon, CR0 4YY

A catalogue record for this title is available from the British Library.

ISBN 978 1 4718 9219 6

CONTENTS

How to get the most from this book

HOW TO GET THE MOST FROM THIS BOOK

Welcome to the CCEA GCSE Single Award Science Student Book.

This book covers all of the content for the 2017 CCEA GCSE Single Award specification.

The following features have been included to help you get the most from this book.

Specification points

Check that you are covering all the required content for your course, with specification references and a brief overview of each chapter.

Show you can

Complete the Show you can tasks to prove that you are confident in your understanding of each topic.

Tip

These highlight important facts, common misconceptions and signpost you towards other relevant chapters. They also offer useful ideas for remembering difficult topics.

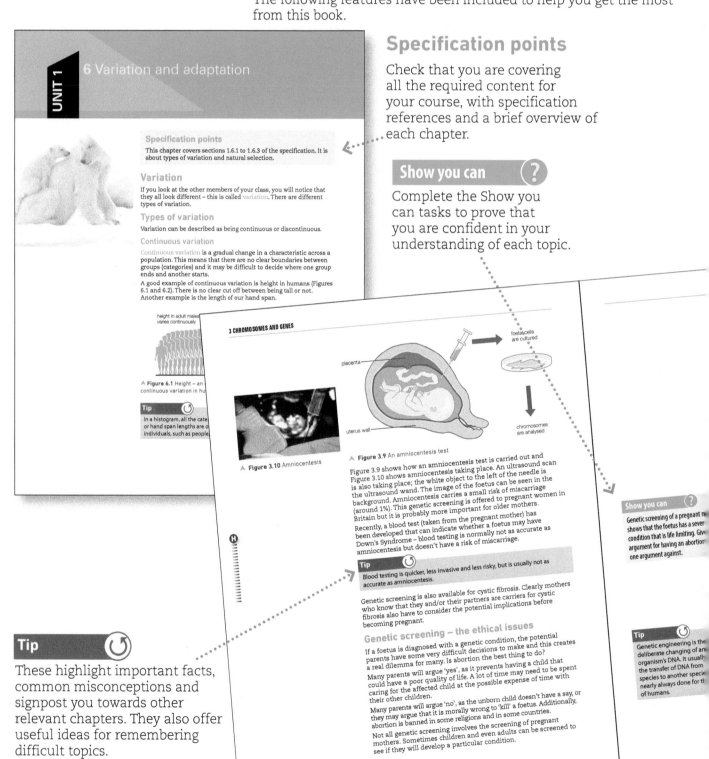

UNIT 1

6 Variation and adaptation

Specification points
This chapter covers sections 1.6.1 to 1.6.3 of the specification. It is about types of variation and natural selection.

Variation
If you look at the other members of your class, you will notice that they all look different – this is called variation. There are different types of variation.

Types of variation
Variation can be described as being continuous or discontinuous.

Continuous variation
Continuous variation is a gradual change in a characteristic across a population. This means that there are no clear boundaries between groups (categories) and it may be difficult to decide where one group ends and another starts.

A good example of continuous variation is height in humans (Figures 6.1 and 6.2). There is no clear cut off between being tall or not. Another example is the length of our hand span.

height in adult males varies continuously

▲ **Figure 6.1** Height – an continuous variation in hu

Tip
In a histogram, all the cate or hand span lengths are o individuals, such as people

---

3 CHROMOSOMES AND GENES

placenta

uterus wall

foetal cells are cultured

chromosomes are analysed

▲ **Figure 3.10** Amniocentesis

▲ **Figure 3.9** An amniocentesis test

Figure 3.9 shows how an amniocentesis test is carried out and Figure 3.10 shows amniocentesis taking place. An ultrasound scan is also taking place; the white object to the left of the needle is the ultrasound wand. The image of the foetus can be seen in the background. Amniocentesis carries a small risk of miscarriage (around 1%). This genetic screening is offered to pregnant women in Britain but it is probably more important for older mothers.

Recently, a blood test (taken from the pregnant mother) has been developed that can indicate whether a foetus may have Down's Syndrome – blood testing is normally not as accurate as amniocentesis but doesn't have a risk of miscarriage.

Tip
Blood testing is quicker, less invasive and less risky, but is usually not as accurate as amniocentesis.

Genetic screening is also available for cystic fibrosis. Clearly mothers who know that they and/or their partners are carriers for cystic fibrosis also have to consider the potential implications before becoming pregnant.

Genetic screening – the ethical issues
If a foetus is diagnosed with a genetic condition, the potential parents have some very difficult decisions to make and this creates a real dilemma for many. Is abortion the best thing to do?

Many parents will argue 'yes', as it prevents having a child that could have a poor quality of life. A lot of time may need to be spent caring for the affected child at the possible expense of time with their other children.

Many parents will argue 'no', as the unborn child doesn't have a say, or they may argue that it is morally wrong to 'kill' a foetus. Additionally, abortion is banned in some religions and in some countries.

Not all genetic screening involves the screening of pregnant mothers. Sometimes children and even adults can be screened to see if they will develop a particular condition.

Show you can
Genetic screening of a pregnant m shows that the foetus has a sever condition that is life limiting. Giv argument for having an abortion one argument against.

Tip
Genetic engineering is the deliberate changing of an organism's DNA. It usually the transfer of DNA from species to another specie nearly always done for th of humans.

Practicals

These practical tasks contain full instructions on apparatus, method and results analysis and will help develop your practical skills.

CCEA's prescribed practicals are clearly highlighted.

Examples

Examples of questions and calculations that feature full workings and sample answers.

Material for Higher tier only is marked with

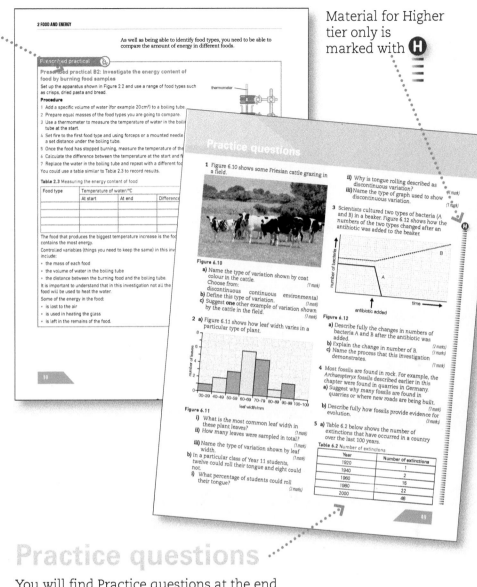

2 FOOD AND ENERGY

As well as being able to identify food types, you need to be able to compare the amount of energy in different foods.

Prescribed practical

Prescribed practical B2: Investigate the energy content of food by burning food samples

Set up the apparatus shown in Figure 2.2 and use a range of food types such as crisps, dried pasta and bread.

Procedure

1 Add a specific volume of water (for example 20 cm³) to a boiling tube.
2 Prepare equal masses of the food types you are going to compare.
3 Use a thermometer to measure the temperature of water in the boiling tube at the start.
4 Set fire to the first food type and using forceps or a mounted needle a set distance under the boiling tube.
5 Once the food has stopped burning, measure the temperature of the
6 Calculate the difference between the temperature at the start and fi
7 Replace the water in the boiling tube and repeat with a different foo
You could use a table similar to Table 2.3 to record results.

Table 2.3 Measuring the energy content of food

Food type	Temperature of water/°C		
	At start	At end	Difference

The food that produces the biggest temperature increase is the foo contains the most energy.

Controlled variables (things you need to keep the same) in this inv include:
• the mass of each food
• the volume of water in the boiling tube
• the distance between the burning food and the boiling tube.
It is important to understand that in this investigation not all the food will be used to heat the water.

Some of the energy in the food:
• is lost to the air
• is used in heating the glass
• is left in the remains of the food.

Practice questions

1 Figure 6.10 shows some Friesian cattle grazing in a field.

Figure 6.10

a) Name the type of variation shown by coat colour in the cattle. *(1 mark)*
Choose from:
discontinuous   continuous   environmental
b) Define this type of variation. *(1 mark)*
c) Suggest **one** other example of variation shown by the cattle in the field. *(1 mark)*

2 a) Figure 6.11 shows how leaf width varies in a particular type of plant.

Figure 6.11

i) What is the most common leaf width in these plant leaves? *(1 mark)*
ii) How many leaves were sampled in total? *(1 mark)*
iii) Name the type of variation shown by leaf width. *(1 mark)*
b) In a particular class of Year 11 students, twelve could roll their tongue and eight could not.
i) What percentage of students could roll their tongue? *(2 marks)*

ii) Why is tongue rolling described as discontinuous variation? *(1 mark)*
iii) Name the type of graph used to show discontinuous variation. *(1 mark)*

3 Scientists cultured two types of bacteria (A and B) in a beaker. Figure 6.12 shows how the numbers of the two types changed after an antibiotic was added to the beaker.

Figure 6.12

a) Describe fully the changes in numbers of bacteria A and B after the antibiotic was added. *(2 marks)*
b) Explain the change in number of B. *(3 marks)*
c) Name the process that this investigation demonstrates. *(1 mark)*

4 Most fossils are found in rock. For example, the Archaeopteryx fossils described earlier in this chapter were found in quarries in Germany.
a) Suggest why many fossils are found in quarries or where new roads are being built. *(1 mark)*
b) Describe fully how fossils provide evidence for evolution. *(3 marks)*

5 a) Table 6.2 below shows the number of extinctions that have occurred in a country over the last 100 years.

Table 6.2 Number of extinctions

Year	Number of extinctions
1920	1
1940	2
1960	15
1980	22
2000	46

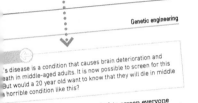

Genetic engineering

's disease is a condition that causes brain deterioration and eath in middle-aged adults. It is now possible to screen for this But would a 20 year old want to know that they will die in middle a horrible condition like this?

ossible (and relatively inexpensive) to screen everyone before birth, in childhood or as an adult) for many different e information obtained is referred to as a genetic profile. is information be available to life insurance companies and s? If insurers had this information they may not provide e, or they may make it more expensive for someone with a ondition.

her thoughts about genetic screening:
costs of screening compared to costs of treating individuals a genetic condition – should cost be a factor?
ld genetic screening be extended to more than just serious tic conditions?
at if it can predict life expectancy?

yourself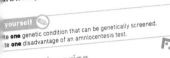
te **one** genetic condition that can be genetically screened.
te **one** disadvantage of an amniocentesis test.

etic engineering

tic engineering is the deliberate modification (changing) of an nism's genome (DNA). While mutations also involve changes to , they are random and almost always unwanted and harmful, reas genetic engineering is carried out in order to benefit nans.

etic engineering involves taking a piece of DNA, usually a gene, one organism (the donor) and adding it to the genetic material nother organism (the recipient). Commonly, DNA for a desired duct (for example a human hormone) is incorporated into the A of bacteria. This is because bacterial DNA is easily manipulated d also because bacteria reproduce so rapidly that large numbers n be produced quickly with the new gene.

a result, the bacteria will produce a valuable product, such as drug or hormone that may be difficult or expensive to produce y other means. Once the new genetic material is added into the acteria, they are allowed to grow and reproduce rapidly in suitable rowing conditions. Special fermenters or bioreactors maximise the roduction of the desired product.

Practice questions

You will find Practice questions at the end of every chapter. These follow the style of the different types of questions you might see in your examination and have marks allocated to each question part.

Test yourself

These short questions, found throughout each chapter, allow you to check your understanding as you progress through a topic.

Answers

Answers for all questions in this book can be found online at:
www.hoddereducation.co.uk/cceagcsesingleaward

1 Cells

Specification points
This chapter covers points 1.1.1 to 1.1.6 of the specification.
It is about using microscopes, animal and plant cells, stem cells
and the specialisation of cells.

Living organisms are made up of microscopic units called cells.
However, the structure of animal cells is different from plant cells.

Animal cells

Figure 1.1 shows a typical animal cell.

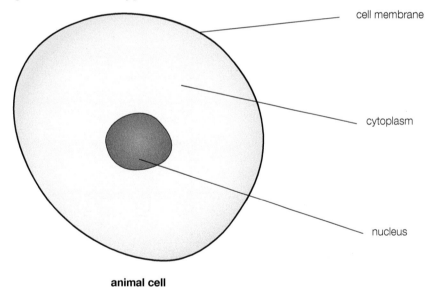

cell membrane

cytoplasm

nucleus

animal cell

⋀ **Figure 1.1** A typical animal cell

Animal cells contain a selectively permeable cell membrane that
forms a boundary to the cell and controls what enters or leaves.
Chemical reactions take place in the cytoplasm. The nucleus is
the control centre of the cell and genetic information is stored in
chromosomes inside it.

Plant cells

Figure 1.2 shows a typical plant cell.

Like animal cells, plant cells have a cell membrane, cytoplasm and
nucleus. However, in addition they also have:

- a cellulose cell wall – a rigid (stiff) structure immediately outside the cell membrane that provides support
- a large permanent vacuole – contains cell sap and when full pushes the cell membrane against the cell wall, making the cell rigid and providing support
- chloroplasts – contain chlorophyll, which traps light and helps the plant make food during photosynthesis. Chloroplasts are not found in all plant cells – they are only found in the green parts of plants, particularly the leaves.

Tip

Chloroplasts are only present in leaf (green) cells so will be absent from cells from other parts of a plant, such as roots.

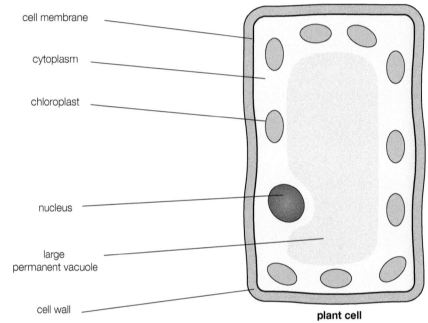

cell membrane

cytoplasm

chloroplast

nucleus

large permanent vacuole

cell wall

plant cell

▲ **Figure 1.2** A typical plant cell

Prescribed practical

Prescribed practical B1: Make a temporary slide and use a light microscope to examine and identify the structures of a typical plant and animal cell

In this practical, you will carry out practical work to make a temporary slide and use a light microscope to examine and identify the structures of a typical plant and animal cell.

In this practical you will use a light microscope to examine the structures of plant and animal cells.

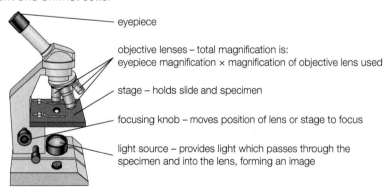

eyepiece

objective lenses – total magnification is:
eyepiece magnification × magnification of objective lens used

stage – holds slide and specimen

focusing knob – moves position of lens or stage to focus

light source – provides light which passes through the specimen and into the lens, forming an image

▲ **Figure 1.3** The light microscope

Using microscopes

When using a microscope it is important to clip the slide containing the cells (the specimen) you want to look at tightly into place on the microscope stage. It is also important to focus at low power first. With low power you have a wider field of view (can see more cells) and it is easier to find what you are looking for. It is also much easier to focus at low power.

After you have observed the specimen at low power you may want to look at it at a higher magnification. Change the objective lens to give you a higher power then refocus.

The magnification of a microscope is the eyepiece lens magnification × the objective lens magnification.

When focusing it is very important that the objective lens does not come into contact with the slide – this is particularly likely to happen at high power. To prevent this you can move the lens down until it is almost touching the slide before you attempt to focus, and then focus as you move the lens and slide further apart.

Making slides of plant cells

Procedure

1 Peel a small section of thin and transparent tissue from the inside of an onion.

2 Using forceps, place the onion tissue evenly on the centre of a microscope slide.

3 Use a drop pipette to add water to the onion tissue to stop it drying out. Your teacher might suggest you use iodine instead of water – this is because iodine stains some parts of the cell, for example the nucleus, making these parts easier to see.

4 Gently lower a coverslip onto the onion tissue. It is better to load the coverslip one end first as this prevents trapping air bubbles. The coverslip will help protect the lens should the lens make contact with the slide and also prevents the cells drying out. Figure 1.4 shows how to do this.

5 Set the slide onto the stage of the microscope and examine it using low power first and then high power.

Example

What is the magnification when there is an eyepiece lens × 10 and an objective lens × 10?

Answer

10 × 10 = 100

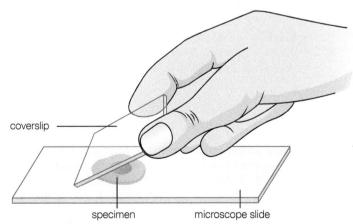

coverslip

specimen   microscope slide

▲ **Figure 1.4** Making a microscope slide

When using thin sections of an onion you should be able to see the cell wall and probably the nucleus in some cells. You are unlikely to see other parts such as the cell membrane (too thin to see) or the chloroplasts (onion cells are taken from the part of the onion that grows underground and therefore will not contain any chloroplasts).

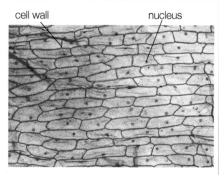

▲ **Figure 1.5** Onion cells as seen under the microscope at high power

Tip

The onion cells in Figure 1.5 have been stained with iodine – this makes the nucleus more obvious. With water the nuclei are often very difficult to see.

Making slides of animal cells

Procedure

1 Using your nail or an inter-tooth brush gently scrape the inside of your cheek.
2 Smear the material gathered onto the centre of a microscope slide. Your teacher may suggest you add a stain such as methylene blue which makes the cells more obvious.
3 Carefully lower a coverslip on top as described before.
4 Observe using a light microscope; first at low power, then using high power.

Note: As with all practical work it is important to be safety aware. Your teacher will tell you how to dispose of the used slides.

Tip

Animal cells are usually much smaller than plant cells, so they may be harder to find on the slide.

Questions

1 Give two reasons why it is important to observe cells at low power first (before using high power).
2 Give two differences between onion cells and animal cells as seen under the microscope.

Show you can

A student was using a microscope to observe some onion cells but the image appeared blurry. Suggest a reason for this and describe how he could fix the problem.

Test yourself

1 State **three** structures that are present in both plant and animal cells.
2 Name **two** structures that are present in plant cells but not in animal cells.
3 What is the function of the nucleus?
4 A microscope has an eyepiece of magnification × 4 and a lower power objective lens of × 10. What is the total magnification at low power?

Stem cells

Stem cells are very simple cells found in animals and plants that can divide to form other cells of the same general type. Animals and plants grow by increasing their number of cells.

Stem cells are only found in certain parts of the body. For example, stem cells can be found in the bone marrow. These bone marrow cells can make the different types of blood cells (but only blood cells). In plants, stem cells are found in the tips of shoots and roots.

Once stem cells become specialised into particular cell types they lose their ability to differentiate into other cell types.

H

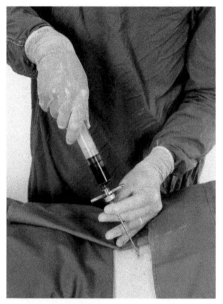

▲ **Figure 1.6** Obtaining bone marrow from a donor before it is transplanted into someone with leukaemia

Stem cells in medicine

Leukaemia is a type of cancer of the blood. Bone marrow transplants can be used as a form of treatment (see Figure 1.6). The stem cells in the bone marrow from a donor have the ability to produce the different types of blood cells in the right proportions (which doesn't happen in a person with leukaemia).

Although stem cell research is an important area of medical research, some people are very concerned, for a number of reasons:

▶ Stem cell research has ethical implications. Some people think that stem cell research using embryo stem cells could lead to 'designer babies'.

▶ There may be a risk of the process going wrong and viruses or diseases being transferred into the patient from the bone marrow cells being used.

▶ There may be a risk that tumours (cancer) or other unwanted cell types will develop.

▶ Before carrying out a bone marrow transplant (with bone stem cells) in the treatment of leukaemia, it is often necessary to destroy the cancerous cells in the patient using radiotherapy or chemotherapy during pre-treatment. This pre-treatment can kill the cancerous cells, but also the healthy cells in the region too. Killing other healthy cells can leave the patient with a very weak immune system and unable to fight off disease.

As a result of the ethical implications, and potential risks involved, the government carefully monitors and regulates research in this area. Most scientists agree that stem cell research will have huge benefits for human health in the coming decades.

Cell specialisation

The function of stem cells in multicellular organisms is to ensure that all the different types of cells needed by these organisms are formed. There are a number of levels of organisation. In animals there are:

▶ cells – the basic building blocks of all living organisms

▶ tissues – groups of cells with similar structures and functions

▶ organs – groups of tissues that have similar functions

▶ organ systems – organs are organised into organ systems

▶ organisms – the different organ systems make up the organisms.

Table 1.1 gives some examples of these levels of organisation.

Table 1.1 Levels of organisation in multicellular animals

Organisational level	Example
cell	nerve cell, red blood cell
tissue	nervous tissue, blood tissue
organ	brain, heart
organ system	nervous system, circulatory system
organism	human, goat

Plants are also organised at different levels. For example, the leaf is an organ containing leaf tissue.

Test yourself

5 Define the term 'stem cell'.
6 Name the organisation level between cell and organ.

Show you can

Give the arguments for and against using stem cells in medicine.

Practice questions

1 Copy and complete the following sentences. Choose from:

cell wall : nucleus : vacuole : cell membrane

Plant and animal cells both have a
.... and a ....

However, only plant cells have a
.... and a........................ *(2 marks)*

2 **a)** You have been given a thin section of onion cells (onion epidermis). Describe how you would make a temporary slide of onion cells suitable for viewing under the microscope.
(3 marks)

b) When observing a specimen on a slide, give **one** reason why it is important to focus at low power before focusing at high power. *(1 mark)*

3 **a)** What is a stem cell? *(1 mark)*
b) Give **two** places where stem cells can be found. *(2 marks)*

4 **a)** **i)** Describe the role of stem cells in bone marrow transplants in the treatment of leukaemia. *(1 mark)*
ii) Explain why a stem cell transplant is needed. *(1 mark)*
b) Give **two** risks associated with using stem cells in bone marrow transplants. *(2 marks)*

2 Food and energy

Specification points

This chapter covers sections 1.2.1 to 1.2.13 of the specification. It is about food and energy, biological molecules, nutrition and food tests, food and health, and exercise and respiration.

Food and diet

We all know we need to eat food, but do we know why? The different types of food have a wide range of different functions. Therefore, we need a wide range of different food types in our diet.

Food types

Table 2.1 shows the main food types (biological molecules) that we need to keep us healthy.

Table 2.1 Food types

Food type	Role in body	Examples
carbohydrate – starch	to provide slow-release energy	potato bread
carbohydrate – sugars	to provide fast-release energy	cake biscuits
protein	for growth and repair of body cells	fish beans
fat	to provide a lot of energy and can be used for insulation	sausages butter
vitamin C	to keep the body in good working order, particularly teeth and gums	oranges lemons
vitamin D	important for normal growth of bones and teeth; sunlight also helps us make vitamin D	fish
calcium (mineral)	important for normal growth of bones and teeth	milk cheese
iron (mineral)	to help the red blood cells carry oxygen	red meat spinach
water	needed as: • a solvent and for reactions that cannot take place without water • a transport medium, for example in blood	
fibre	needed to prevent constipation	wholemeal bread green vegetables

Food tests

Some food types can be identified by carrying out food tests. The tests you need to know about are described in Table 2.2.

Table 2.2 Food tests

Food type	Reagent	Method	Initial colour	End colour if type present
sugar	Benedict's	add Benedict's reagent to the food sample and heat in a water bath	blue	brick red precipitate
starch	iodine	add iodine	yellow-brown	blue-black
protein	Biuret	add sodium hydroxide, then a few drops of copper sulfate and shake	blue	lilac/purple
fat	ethanol	shake the fat with ethanol in a boiling tube, then add an equal volume of water	clear	white emulsion

Tip

Remember that these food tests show which food types are present but they do not really show how much is present. The Benedict's test is an exception as if there is only a small amount of sugar present the colour will change to green or orange (but not brick red).

Tip

Remember that an unknown food being tested could contain more than one food type. For example, bacon will contain both protein and fat.

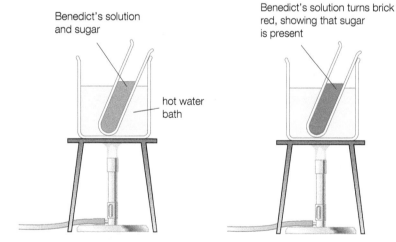

Benedict's solution and sugar

hot water bath

Benedict's solution turns brick red, showing that sugar is present

▲ **Figure 2.1** Testing with Benedict's solution

You should be able to carry out a food test on a food sample to find out which food types are present using one or more of the tests in Table 2.2.

Show you can (?)

Use Tables 2.1 and 2.2 to describe what will happen if you test a potato chip with iodine.

Test yourself ✎

1 Name the food types (biological molecules) that provide energy.
2 Fish is a good source of two food types, name both.
3 Name the food tests (reagents) that test for different types of carbohydrate.

As well as being able to identify food types, you need to be able to compare the amount of energy in different foods.

Prescribed practical B2: Investigate the energy content of food by burning food samples

Set up the apparatus shown in Figure 2.2 and use a range of food types such as crisps, dried pasta and bread.

Procedure

1 Add a specific volume of water (for example 20 cm³) to a boiling tube.
2 Prepare equal masses of the food types you are going to compare.
3 Use a thermometer to measure the temperature of water in the boiling tube at the start.
4 Set fire to the first food type and using forceps or a mounted needle hold it a set distance under the boiling tube.
5 Once the food has stopped burning, measure the temperature of the water.
6 Calculate the difference between the temperature at the start and finish.
7 Replace the water in the boiling tube and repeat with a different food.

You could use a table similar to Table 2.3 to record results.

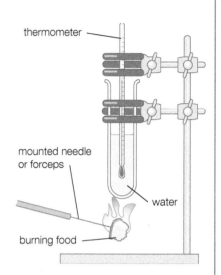

▲ **Figure 2.2** Measuring the energy content of food

Table 2.3 Measuring the energy content of food

Food type	Temperature of water/°C		
	At start	At end	Difference

The food that produces the biggest temperature increase is the food that contains the most energy.

Controlled variables (things you need to keep the same) in this investigation include:

- the mass of each food
- the volume of water in the boiling tube
- the distance between the burning food and the boiling tube.

It is important to understand that in this investigation not all the energy in the food will be used to heat the water.

Some of the energy in the food:

- is lost to the air
- is used in heating the glass
- is left in the remains of the food.

Questions and sample data

1 In an investigation comparing the amount of energy in food, a student obtained the following results.

Table 2.4 Testing food samples

Food sample	Temperature of water/°C		
	At start	At end	Difference
A	18	26	8
B	18	31	13

a) Which food sample (A or B) contained more energy?

b) One of the foods was bacon and the other one was bread. Suggest which food was bacon. Explain your answer.

c) State one thing that should have been kept the same in this investigation (controlled variable).

How much energy do we need?

Different people need different amounts of energy.

age
teenagers usually need more energy than elderly people
activity
active people need more energy than less active people
factors that affect our energy needs
gender
boys often need more energy than girls

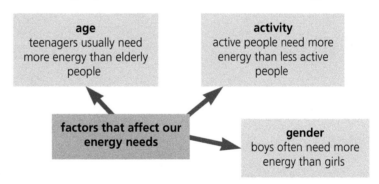

Figure 2.3 Factors that affect energy needs

You should understand that the factors are often interlinked. For example, teenagers usually need more energy than elderly people because they are more active.

There are exceptions; for example, some girls may be more active than some of the boys in the same class and therefore they need more energy.

Food and health

A poor diet can lead to ill health. Coronary heart disease (CHD), which kills many people in Northern Ireland, is often linked to a poor diet.

How heart attacks happen

cholesterol (a fatty substance) builds up on the coronary artery walls
the blood flow to the heart becomes blocked
oxygen and glucose cannot reach the heart muscle
the heart muscle dies as it cannot respire and produce energy

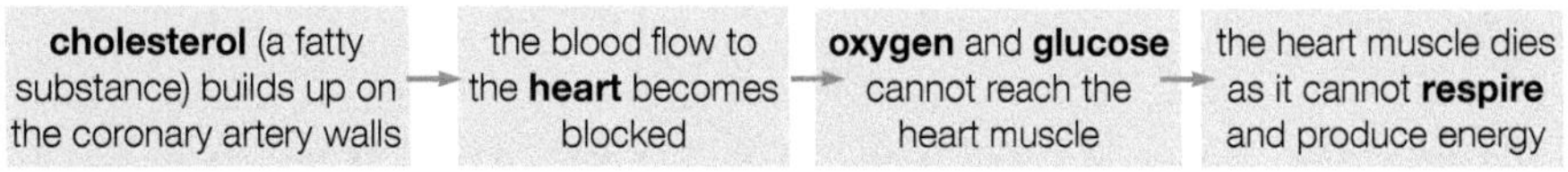

Figure 2.4 Flow chart showing how heart attacks happen

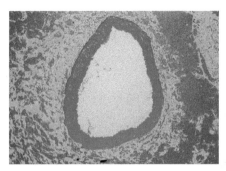

⋏ **Figure 2.5a** A healthy coronary artery

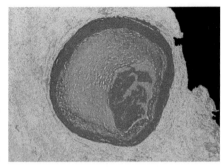

⋏ **Fig 2.5b** A coronary artery that is almost blocked and at risk of causing a heart attack

There are many factors that contribute to heart disease. There are therefore many things we can do to reduce our chances of getting heart disease. These factors can be grouped into lifestyle factors and dietary factors.

Tip

Many students mix up dietary factors and lifestyle factors in exams. In Single Award Science, diet is not classified as a lifestyle factor.

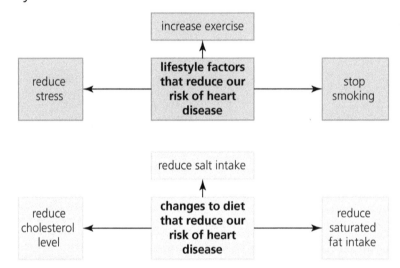

⋏ **Figure 2.6** Lifestyle and dietary factors that can reduce the risk of heart disease

As well as the factors listed above, it is very important to avoid obesity if possible. Being obese increases the risk of poor health. However, it is very hard to convince some people that there is a link between diet, lifestyle and heart disease.

Strokes affect the brain. People who have strokes may become partially paralysed as parts of the brain stop working. The sections above show that there is a link between CHD, lifestyle and poor diet. These links exist for the development of strokes too. In general, positive lifestyle changes and dietary changes also protect against strokes.

The costs of circulatory diseases to society

Strokes and heart disease are both circulatory diseases. They affect much more than just the patient involved.

Whole families are affected as patients are often very ill and need a lot of care. Heart disease and stroke patients are often unable to work for a long time.

Heart disease and strokes are very expensive to treat because:

▶ patients are often in hospital for a long time

▶ expensive drugs and medicines are often needed

▶ many highly-trained staff are needed to care for the patients.

The effect of exercise on heart rate and recovery rate

We can help our heart by taking regular exercise. The graph in Figure 2.7 shows the effect of exercise on pulse (heart) rate and recovery rate.

The recovery rate is the time it takes for the pulse or heart rate to return to normal. This will usually be shorter for people who exercise regularly or play a lot of sport.

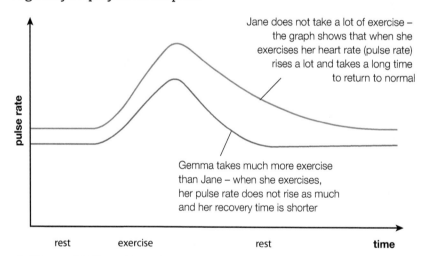

Jane does not take a lot of exercise – the graph shows that when she exercises her heart rate (pulse rate) rises a lot and takes a long time to return to normal

Gemma takes much more exercise than Jane – when she exercises, her pulse rate does not rise as much and her recovery time is shorter

▲ **Figure 2.7** The effect of exercise on heart rate

Tip

Graphs (or tables) showing the effect of exercise on pulse rate (heart rate) are common in exams. As well as being able to interpret the graphs, you could be asked to calculate the increase (or decrease) in rate or even the % change. Remember that % change is the change in pulse rate divided by the value at the start multiplied by 100.

Jane had a resting heart rate of 70 bpm. She exercised for 5 minutes and her heart rate had increased to 98 bpm by the time she finished exercising. What was the percentage increase in Jane's heart rate?

Answer

Increase was

98 − 70 = 28

Percentage increase was

$\frac{28}{70} \times 100 = 40\%$

The effect of exercise on the heart

Regular exercise also strengthens the heart muscle (as it does any muscle).

A stronger heart muscle can increase the cardiac (heart) output (even when not exercising) – this means the heart can pump more blood in each beat and therefore more each minute. An advantage of this is that the heart has to beat less often to pump the same amount of blood. If the heart pumps less often it will suffer less wear and tear.

Respiration

Carbohydrates and fats in the body provide chemical energy. The energy in these foods can be converted into a form of energy that the body can use by the process of respiration.

The word equation for respiration is:

glucose + oxygen → carbon dioxide + water + energy

Respiration is an exothermic reaction that releases energy. This means that heat energy is given off during the reaction. In mammals, it is the heat given off that gives us our high body temperature.

The balanced symbol equation for respiration is:

$C_6H_{12}O_6 + 6O_2 \rightarrow 6CO_2 + 6H_2O + energy$

4 State **two** lifestyle factors that can help protect against CHD.

5 Why is it important to reduce the amount of saturated fat in the diet?

6 Give **one** reason why we need to have fat in our diet.

7 Why do we need to respire?

Explain why our heart rate increases when we exercise. To answer this, you should think about the relationship between the role of respiration in the body and the role of the blood in transporting oxygen and glucose to body cells.

Practice questions

1 a) Copy and complete Table 2.5 about food tests.

(3 marks)

Table 2.5

Name of test	Food type	Result if food is present
iodine		turns from yellow-brown to blue-black
Benedict's	sugar	
	protein	turns from blue to lilac

b) Name the food test that requires heating. *(1 mark)*

2 a) Table 2.6 shows the results produced when a student compared the amount of energy in three foods (A, B and C).

Table 2.6

Food	Starting temperature/°C	Final temperature/°C	Temperature rise/°C
A	16	24	8
B	16	28	
C	16	35	19

    i) Copy and complete the table to show the temperature rise for B. *(1 mark)*

    ii) Which food (A, B or C) contained the most energy? *(1 mark)*

b) State **two** things that the student would need to have done to obtain valid results (keep the test fair). *(2 marks)*

c) The temperature increases shown in the table are probably an underestimation of the actual amount of energy in the food. Give **one** reason for this. *(1 mark)*

3 a) We can reduce our chances of having heart disease by making certain lifestyle changes. Which of the following changes is a lifestyle change that can reduce our risk? *(1 mark)*

    reduce saturated fat in the diet

    increase stress

    stop smoking

b) In which part of the body do strokes occur? *(1 mark)*

c) Give **one** reason why the cost of treating heart disease is so expensive. *(1 mark)*

4 a) The graph in Figure 2.8 shows the effect of exercise on the pulse rate of two girls.

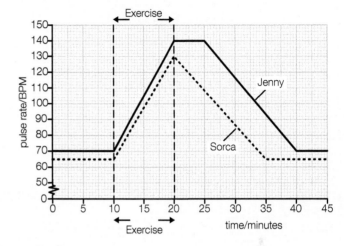

▲ **Figure 2.8**

    i) Calculate the increase in Sorca's pulse rate during exercise. *(1 mark)*

    ii) Calculate the percentage increase in Sorca's pulse rate during exercise. *(1 mark)*

b) Use the graph to suggest which girl exercises more regularly. Explain your choice giving **two** pieces of evidence from the graph. *(2 marks)*

c) If the girls continued to exercise for a period of time, suggest the effect this would have, if any, on recovery time. *(1 mark)*

5 Exercise helps strengthen the heart.

a) What is meant by the term cardiac output? *(1 mark)*

b) i) State the effect of a stronger heart on cardiac output. *(1 mark)*

    ii) How does this benefit the heart? *(2 marks)*

3 Chromosomes and genes

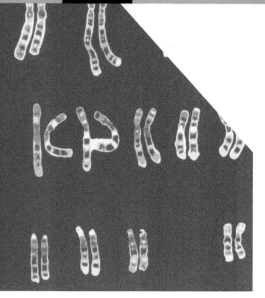

Chromosomes, genes, DNA and the genome

Most living cells contain a nucleus. The nucleus is the control centre of a cell because it contains chromosomes that are divided into smaller sections called genes. The chromosomes are arranged in pairs – humans have 46 chromosomes in each cell arranged in 23 pairs.

The genes in our body control characteristics such as eye and hair colour. In fact, they control all the features that make us what we are.

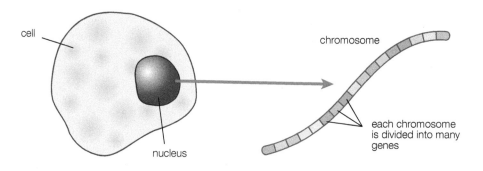

▲ **Figure 3.1** Chromosomes and genes

Chromosomes and genes are made of DNA (deoxyribose nucleic acid). This is a very important chemical with special properties. The DNA molecule is formed in a double helix in which the two strands are linked as shown in Figure 3.2.

Tip ↻

DNA is the name of the molecule (the core component) that makes genes and chromosomes; the double helix refers to its shape or the way it is structured.

▲ **Figure 3.2** A section of the DNA double helix

Tip

You should be able to identify genes as sections of chromosomes made up of short lengths of DNA that operate as functional units to control characteristics.

Tip

In humans, the genome is all the DNA in all 46 chromosomes.

All the DNA in an organism (all the genetic material) is referred to as its genome.

Mutations

Mutations are random changes in the structure or numbers of chromosomes or genes. Although variation is a normal feature of living organisms, the change caused by some mutations can be harmful.

Cancer

When an organism grows, this usually involves it producing more cells. This involves cells dividing (splitting) into two and then each of the two new cells growing to normal size and eventually dividing again and so on.

Normally this cell division is controlled, with only certain cells dividing and cells only dividing a certain number of times. Cancer is uncontrolled cell division caused by damage (change) to the genes or chromosomes in a part of the body. Cancer is caused by mutations.

Some mutations (cancers) can be triggered by environmental factors. Ultraviolet (UV) light coming from the Sun can cause mutations in skin cells leading to skin cancer.

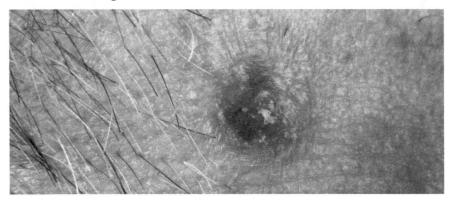

▲ **Figure 3.3** Skin cancer

Although getting a tan can make us feel better and increases the amount of vitamin D we have, it is important that we are careful when exposed to UV light. Things we can do to reduce the damage caused by sunlight include:

▶ using sunscreen
▶ wearing a hat to protect the face, eyes and neck
▶ limiting the amount of time in the Sun, particularly when it is strongest (around midday and early afternoon).

Tip

In the UK the number of people with skin cancer is increasing. People going abroad on holiday to get a suntan is a major reason for this.

Genetic conditions

Some medical conditions are caused by problems with chromosomes or genes. Two examples are Down's Syndrome and cystic fibrosis.

▶ Down's Syndrome – caused by a mutation involving chromosome number. Individuals with Down's Syndrome have 47 chromosomes instead of the normal 46. It is possible to tell if someone has Down's Syndrome by studying a karyotype. This is a diagram of someone's chromosomes laid out so they can be counted (Figure 3.4).

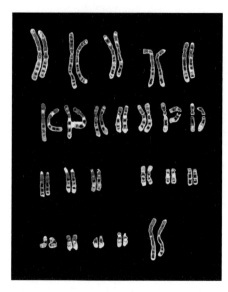
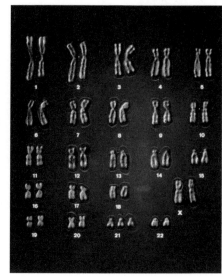

 Figure 3.4 Karyotypes of a normal individual (left) and an individual with Down's Syndrome (right) – note that with Down syndrome there is an extra chromosome 21

▶ Cystic fibrosis – unlike Down's Syndrome, individuals have cystic fibrosis due to gene mutations being inherited from parents at birth. It is an example of an inherited disease.

Show you can (?)

Explain how sunlight can lead to skin cancer.

Test yourself ✎

1 Place the structures below in order of size, starting with the largest.
   gene   chromosome   cell   nucleus
2 What are mutations?
3 How many chromosomes are there in each cell of a person with Down's Syndrome?

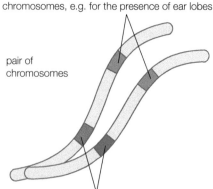

the form of gene (allele) is the same in both chromosomes, e.g. for the presence of ear lobes

pair of chromosomes

the alleles of this gene are different, e.g. one for brown eyes and one for blue eyes

 Figure 3.5 Chromosomes and genes

Genetic diagrams and terminology

The science of genetics explains how characteristics pass from parents to offspring.

Figure 3.5 shows that:

▶ Chromosomes are arranged in pairs – humans have 23 pairs, which is 46 chromosomes in total.

▶ Genes are sections of the chromosome that carry the code for particular characteristics such as eye colour.

▶ Similar genes occupy the same position on both chromosomes in a pair.

▶ Genes exist in different forms, called alleles, and the alleles may be homozygous (the same) or heterozygous (different) on the two chromosomes of a pair.

Some key genetic terms are summarised in Table 3.1.

Table 3.1 Key genetic terms 1

Term	Definition	Example
gene	short section of chromosome that carries code for a particular characteristic	gene for eye colour
allele	a particular form of a gene	brown eyes and blue eyes are due to different alleles of the eye colour gene
homozygous	both alleles of a gene are the same	both alleles are for brown eyes
heterozygous	the alleles of a gene are different	one allele is for brown eyes and the other is for blue eyes – see Figure 3.5

Questions about genetics normally ask you to work out which offspring or children would result from particular parents. Sometimes you may be asked to work backwards and work out the parents.

Genetic diagrams (cross diagrams)

When you are asked to work out the offspring produced from, for example, two heterozygous parents, it is useful to set out a diagram called a genetic diagram or a cross. Figure 3.6 shows a cross using the example of height in peas. Peas can be either tall or short and this is controlled by a single gene that has a tall allele and a short allele. In this example, the gene is for height, and tall and short are the two alleles of the height gene.

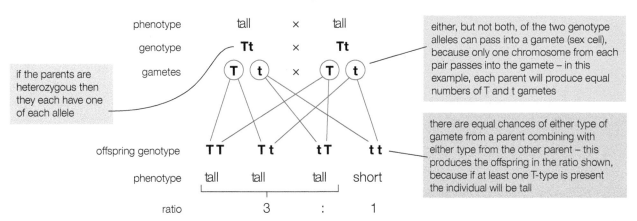

if the parents are heterozygous then they each have one of each allele

either, but not both, of the two genotype alleles can pass into a gamete (sex cell), because only one chromosome from each pair passes into the gamete – in this example, each parent will produce equal numbers of T and t gametes

there are equal chances of either type of gamete from a parent combining with either type from the other parent – this produces the offspring in the ratio shown, because if at least one T-type is present the individual will be tall

Figure 3.6 Tall and short peas

Table 3.2 explains some new terms that you will need to know. Some of these important terms are used in Figure 3.6.

Table 3.2 Key genetic terms 2

Term	Definition	Example
gamete	sex cell that contains only one chromosome from each pair	sperm or egg
genotype	the paired symbols showing the allele arrangement in an individual	the parents in the cross in Figure 3.6 both have the genotype Tt
phenotype	the outward appearance of an individual	the parents in the example have a tall phenotype
dominant	in the heterozygous condition the dominant allele will override the non-dominant (recessive) allele	the parents in the example are both tall even though they are heterozygous and have a short allele
recessive	the recessive allele will be dominated by the dominant allele – it will only show itself in the phenotype if there are two recessive alleles	only ¼ of the offspring in the cross are short, as only ¼ have no dominant T allele present (Figure 3.6)

Tip

If a condition is described as being caused by a recessive allele, this means that to actually have the condition an individual must have two recessive alleles for that gene.

Tip

In genetic crosses, you could be asked to predict the probability of the offspring having a particular condition. If the chance is 1 in 4 (one of the ratios you came across earlier), then this could be written as 1 in 4; 1:3; 1/4 or 25%.

The cross in Figure 3.7 shows how to use a Punnett square. This is a way of setting out a genetic cross in table format. In this example, using height in peas as before, a heterozygous pea (Tt) is crossed with a homozygous recessive pea (tt).

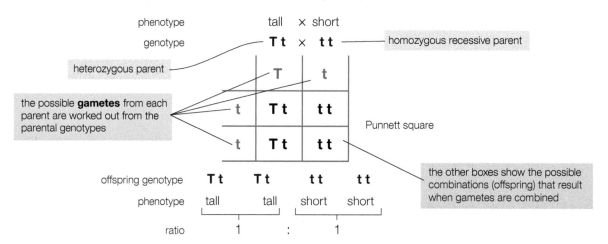

▲ **Figure 3.7** A Punnett square for tall and short peas

Pedigree diagrams

A pedigree diagram shows the way in which a genetic condition is inherited in a family or group of biologically related people. Figure 3.8 is an example of a pedigree diagram showing how the condition albinism is inherited. Albinism is a condition caused by having two recessive alleles present.

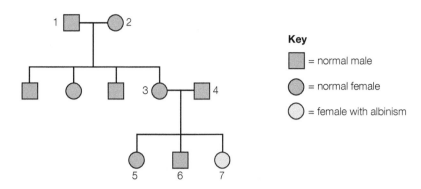

Key

■ = normal male

● = normal female

○ = female with albinism

▲ **Figure 3.8** Pedigree diagram showing inheritance of albinism

In Figure 3.8 one of the grandchildren (7) has albinism. It is possible to use the information provided to work out the probability of other children having the condition.

Show you can ?

Use Figure 3.8 and your knowledge to answer the following questions.
Let A = normal allele; a = albinism allele

a) What is the genotype of the child (7) with albinism?
b) What are the genotypes of the parents of child 7 (3 and 4)?
c) What are the possible genotypes for the brother and sister of the child with albinism (5 and 6)? Explain your answer.
d) What is the probability that the next child of these parents will be a child with albinism?
e) What can you say about the genotypes of the grandparents of child 7?

Genetic counsellors often construct pedigree diagrams and use them to advise parents who have a genetic condition or who may be carriers (heterozygous). Pedigree diagrams can be used in any type of genetic cross but they are obviously very valuable in tracing and predicting harmful genetic conditions.

Genetic screening

Genetic screening may be used to reduce the incidence of diseases or conditions caused by problems with our chromosomes or genes. It involves testing people for the presence of a particular allele or genetic condition. Whole populations can be tested, or targeted groups where the probability of having (or passing on) a particular condition is high. Genetic screening can be a particular issue for pregnant mothers and their partners.

Genetic screening has been available for a long time for Down's Syndrome. In screening for Down's Syndrome, cells can be taken from the amniotic fluid surrounding the baby in the uterus and allowed to multiply in laboratory conditions. The chromosomes in the cells can then be examined to see if the developing foetus has the condition (an amniocentesis test).

Tip

Genetic screening involves checking someone's DNA/genes/chromosomes for the presence of a harmful allele or other problem that may lead to the development of a harmful genetic condition.

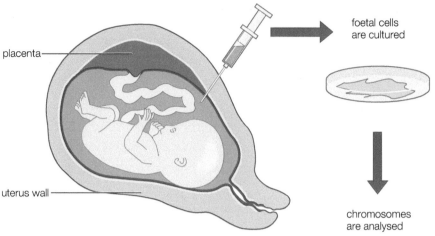

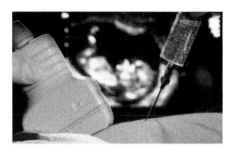

▲ **Figure 3.10** Amniocentesis

▲ **Figure 3.9** An amniocentesis test

Figure 3.9 shows how an amniocentesis test is carried out and Figure 3.10 shows amniocentesis taking place. An ultrasound scan is also taking place; the white object to the left of the needle is the ultrasound wand. The image of the foetus can be seen in the background. Amniocentesis carries a small risk of miscarriage (around 1%). This genetic screening is offered to pregnant women in Britain but it is probably more important for older mothers.

Recently, a blood test (taken from the pregnant mother) has been developed that can indicate whether a foetus may have Down's Syndrome – blood testing is normally not as accurate as amniocentesis but doesn't have a risk of miscarriage.

> **Tip**
>
> Blood testing is quicker, less invasive and less risky, but is usually not as accurate as amniocentesis.

Genetic screening is also available for cystic fibrosis. Clearly mothers who know that they and/or their partners are carriers for cystic fibrosis also have to consider the potential implications before becoming pregnant.

Genetic screening – the ethical issues

If a foetus is diagnosed with a genetic condition, the potential parents have some very difficult decisions to make and this creates a real dilemma for many. Is abortion the best thing to do?

Many parents will argue 'yes', as it prevents having a child that could have a poor quality of life. A lot of time may need to be spent caring for the affected child at the possible expense of time with their other children.

Many parents will argue 'no', as the unborn child doesn't have a say, or they may argue that it is morally wrong to 'kill' a foetus. Additionally, abortion is banned in some religions and in some countries.

Not all genetic screening involves the screening of pregnant mothers. Sometimes children and even adults can be screened to see if they will develop a particular condition.

Example

Huntington's disease is a condition that causes brain deterioration and eventual death in middle-aged adults. It is now possible to screen for this condition. But would a 20 year old want to know that they will die in middle age from a horrible condition like this?

It is now possible (and relatively inexpensive) to screen everyone (whether before birth, in childhood or as an adult) for many different alleles. The information obtained is referred to as a genetic profile. Should this information be available to life insurance companies and employers? If insurers had this information they may not provide insurance, or they may make it more expensive for someone with a genetic condition.

Some other thoughts about genetic screening:

▶ The costs of screening compared to costs of treating individuals with a genetic condition – should cost be a factor?

▶ Should genetic screening be extended to more than just serious genetic conditions?

▶ What if it can predict life expectancy?

Show you can

Genetic screening of a pregnant mother shows that the foetus has a severe condition that is life limiting. Give one argument for having an abortion and one argument against.

Test yourself

7 State **one** genetic condition that can be genetically screened.
8 State **one** disadvantage of an amniocentesis test.

Genetic engineering

Genetic engineering is the deliberate modification (changing) of an organism's genome (DNA). While mutations also involve changes to DNA, they are random and almost always unwanted and harmful, whereas genetic engineering is carried out in order to benefit humans.

Genetic engineering involves taking a piece of DNA, usually a gene, from one organism (the donor) and adding it to the genetic material of another organism (the recipient). Commonly, DNA for a desired product (for example a human hormone) is incorporated into the DNA of bacteria. This is because bacterial DNA is easily manipulated and also because bacteria reproduce so rapidly that large numbers can be produced quickly with the new gene.

As a result, the bacteria will produce a valuable product, such as a drug or hormone that may be difficult or expensive to produce by other means. Once the new genetic material is added into the bacteria, they are allowed to grow and reproduce rapidly in suitable growing conditions. Special fermenters or bioreactors maximise the production of the desired product.

Tip

Genetic engineering is the deliberate changing of an organism's DNA. It usually involves the transfer of DNA from one species to another species and is nearly always done for the benefit of humans.

One of the best examples of genetic engineering providing essential products for humans is the production of genetically engineered human insulin. Diabetes is becoming increasingly common and as a result many more people require insulin than in the past. Before the development of genetic engineering, the insulin was obtained from the pancreases of pigs and cattle. This was a slow and time-consuming process that carried the risk of transmitting diseases. The animal insulin obtained also had small chemical differences from human insulin.

▲ **Figure 3.11** Making human insulin in giant bioreactors

Although most people can see the benefits of genetically engineered insulin, the wider use of genetic engineering can create moral and ethical issues. Theoretically, bacteria or other organisms with new genomes could spread uncontrolled into the wild (the wider environment) and could have unforeseen outcomes that no one could possibly predict.

Practice questions

1 Figure 3.12 represents an animal cell with one chromosome shown.

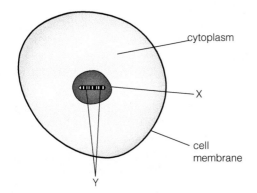

cytoplasm

X

cell membrane

Y

⌃ **Figure 3.12**

a) Name the part of the cell labelled X. *(1 mark)*

b) Name the parts of the chromosome labelled Y. *(1 mark)*

c) Chromosomes also contain DNA. What name is given to the shape of DNA? *(1 mark)*
Choose from:

spiral : double helix : single helix

2 a) Copy and complete the following sentences. Choose from:

mutations : genome : inherited : chromosomes
All the DNA in an organism is called its
.... The DNA is found in long thread-like structures called ....
Changes in DNA are called .... *(3 marks)*

b) Copy and complete Table 3.3 about genetic conditions.

Table 3.3

Genetic condition	Type of mutation	Number of chromosomes present	Inherited (passed from parents to child)
Cystic fibrosis	gene	46	
Down syndrome			No

(3 marks)

3 Cystic fibrosis is caused by a recessive allele.

a) Describe what is meant by the term 'recessive'. *(1 mark)*

b) It is possible for parents who do not have cystic fibrosis to have a child with cystic fibrosis.

i) Copy and complete the Punnett square to show how two heterozygous parents could have a child with cystic fibrosis.
Let F = no CF; f = CF *(2 marks)*

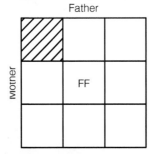

Father

Mother

FF

⌃ **Figure 3.13**

ii) How many different genotypes are shown in the Punnett square? *(1 mark)*

iii) What is the probability of the next child of these parents having cystic fibrosis? *(1 mark)*

4 Presence or absence of a widow's peak in humans is genetically determined by the two alleles of one gene.

⌃ **Figure 3.14**

The allele for having a widow's peak is dominant.
Let W = widow's peak; w = no widow's peak

a) State the genotype(s) that will result in the presence of a widow's peak. *(1 mark)*

b) Copy and complete the Punnett square to show the possible genotypes of children from parents, one of whom is heterozygous and one of whom is homozygous recessive. *(2 marks)*

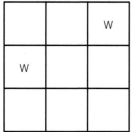

▲ **Figure 3.15**

c) What proportion of the offspring are homozygous? *(1 mark)*

d) What is the probability of this couple's next child having a widow's peak? *(1 mark)*

5 a) Figure 3.16 shows how a mother's age affects the risk of having a child with Down's Syndrome.

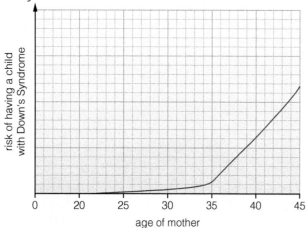

▲ **Figure 3.16**

i) A test can be carried out on foetal cells extracted from the uterus to see if a foetus has Down's Syndrome. Name this test. *(1 mark)*

ii) Using Figure 3.16 and your knowledge, give **two** reasons why this test is normally only offered to pregnant women aged 35 or over. *(2 marks)*

b) It is now possible to obtain an adult's overall genetic profile which can predict the likelihood of having certain illnesses later in life.

i) Suggest **one** reason why many people would not want this information to become public knowledge. *(1 mark)*

ii) Suggest **one** advantage in making this type of information available to health professionals. *(1 mark)*

4 Co-ordination and control

The nervous system

We are able to respond to the environment around us. Anything that we respond to is called a stimulus.

In animals, each type of stimulus affects a receptor in the body. There are many types of receptor, each responding to a particular type of stimulus. If a receptor is stimulated, it may cause an effector such as a muscle to produce a response.

| stimulus | → | receptor | → | effector | → | response |

▲ **Figure 4.1** Flow chart showing a response to a stimulus

The flow chart above is a simplification as it suggests that any stimulation will automatically produce a response. In reality, if we hear a sound (the stimulus), we might respond or not, depending on what the sound is.

Receptors are often found in sense organs. The main senses and their sense organs are listed in Table 4.1.

Table 4.1 Senses and sense organs

Sense	Sense organ
smell	nose
touch	skin
sight	eye
sound	ear
taste	tongue

Coordination

In reality, the receptors and effectors (muscles) are linked by a coordinator. This is usually the brain but may also be the spinal cord. Together, these two structures make up the Central Nervous System (CNS).

Nerve cells or neurones link the receptors and effectors to the coordinator. A neurone carries information in the form of small electrical charges called nerve impulses.

The CNS acts as a 'linking system' and determines which receptors link up with which effectors, and even whether or not a particular stimulus brings about a response.

A more complete flow diagram than the previous one is shown in Figure 4.2.

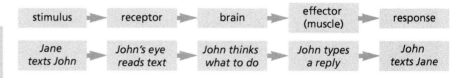

stimulus	→	receptor	→	brain	→	effector (muscle)	→	response
Jane texts John	→	John's eye reads text	→	John thinks what to do	→	John types a reply	→	John texts Jane

▲ **Figure 4.2** A detailed diagram of a response to a stimulus

Voluntary and reflex actions

Voluntary actions are those actions that are deliberate and involve the process of 'thinking'. We choose to do voluntary actions – they are not automatic.

Reflex actions can be best explained by giving an example. If we accidentally touch a very hot object, we respond immediately by rapidly withdrawing our hand from the danger area. The advantage of this is that the hand is moved away before it can get burned too badly. This type of action does not involve any 'thinking' time, as the time taken to consider a response would cause unnecessary damage to the body. This type of action is typical of reflex actions.

All reflex actions have three main characteristics in common as they:
▶ occur very rapidly
▶ are automatic
▶ do not involve conscious control (thinking time).

Reflexes occur very rapidly as the nerve pathway taken is the shortest possible between the receptor and the muscle involved. In a reflex pathway, the total length of the nerve pathway is kept as short as it possibly can be. For example, the knee jerk reflex travels from the knee up to the base of the spinal cord and back into the leg (and does not involve 'thinking' time in the brain).

The reflex arc

A reflex arc is the nerve pathway involved in a reflex. It usually involves three neurones; a sensory, association and motor neurone.
▶ sensory neurone – carries impulses from a receptor to the CNS
▶ association neurone – connects the sensory neurone with a motor neurone
▶ motor neurone – carries impulses from the CNS to an effector (muscle).

Reflex arcs take the shortest possible route between receptor and effector. In addition, there are relatively few gaps between neurones (synapses), as in these gaps the impulses travel relatively slowly. Synapses are important as these can act as junctions controlling which neurones pass on impulses to which other neurones.

Figure 4.3 shows the nerve pathway involved when a hand touches a hot object. There are three types of neurone involved in this response.

Tip

The CNS (brain and spinal cord) ensures that nerve impulses (electrical signals) get from the right receptor (such as a sense organ) to the right effector (such as a muscle).

Tip

In general, the nervous system works very rapidly. Reflex actions are even faster than normal voluntary responses.

Tip

Many reflexes are protective – for example, moving your hand away from a hot object or coughing if something gets stuck in your throat.

Tip

The association neurone is the only neurone that lies totally within the CNS.

Tip

Remember that the neurones in a reflex arc are three different types of specialised nerve cell, each with a cell body and a long extension that carries the nerve impulses. Neurones are very specialised animal cells.

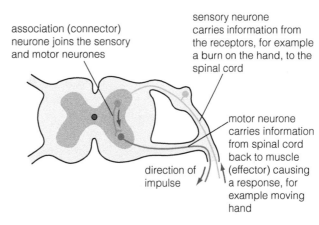

association (connector) neurone joins the sensory and motor neurones

sensory neurone carries information from the receptors, for example a burn on the hand, to the spinal cord

motor neurone carries information from spinal cord back to muscle (effector) causing a response, for example moving hand

direction of impulse

▲ **Figure 4.3** The reflex arc

The diagram shows that both the association and motor neurones begin with the cell body (unlike the sensory neurone). The diagram also shows that only two synapses (the short gaps between neurones) are involved in this pathway.

Tip

Most effectors are muscles but some are also glands that produce hormones.

Show you can

Explain how a reflex arc is adapted to ensure that reflex actions are so rapid.

Test yourself

1 What is a reflex arc?
2 Which neurone brings an impulse away from the CNS to the effector?

Hormones

Another type of messenger system used by the body to bring about responses involves the use of special chemicals called hormones.

Hormones are chemical messengers produced by glands (special structures in the body) that release them into the blood. Although the hormones travel all around the body in the blood, they only affect certain organs, called target organs. The target organs differ for each hormone, although some hormones affect many organs.

Hormones usually act more slowly than the nervous system and act over a longer period of time.

Summarising the differences between the nervous system and hormones

Figure 4.4 shows the differences between the nervous system and hormones.

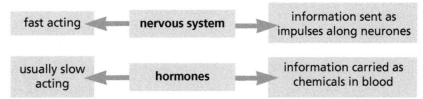

| fast acting | ← **nervous system** → | information sent as impulses along neurones |

| usually slow acting | ← **hormones** → | information carried as chemicals in blood |

▲ **Figure 4.4** Hormones and the nervous system

Insulin

A very important hormone you need to know about is insulin.

It is vital to keep the amount of glucose (sugar) in the blood at just the right level. If it gets too high it can cause the body harm.

Figure 4.5 shows how insulin stops blood glucose levels from rising too high.

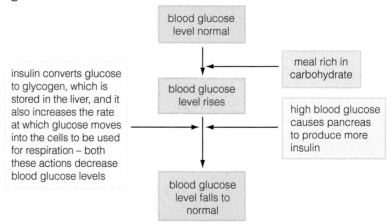

▲ **Figure 4.5** The action of insulin

▲ **Figure 4.6** The effect of insulin on blood glucose levels

Diabetes

Diabetes is a condition in which the blood glucose control mechanism fails.

The symptoms (what happens if the diabetes is untreated) of diabetes include:

▶ high blood glucose levels

▶ glucose in the urine

▶ lethargy (no energy)

▶ thirst.

Types of diabetes

Type 1 diabetes is the form of the condition where the pancreas stops producing insulin. Type 1 diabetes usually develops in children or young adults. People with Type 1 diabetes have to take insulin for the rest of their life.

▲ **Figure 4.7** This girl has type 1 diabetes and is injecting herself with insulin

Tip

Patients with both types of diabetes have to monitor their blood glucose levels regularly so that they know how much medication they need to take.

Tip

Type 2 diabetes is linked to lifestyle but type 1 diabetes is not caused by lifestyle.

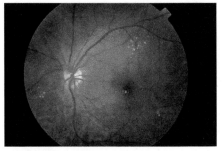

▲ **Figure 4.8** The retina of a diabetes sufferer – the areas of small yellow dots, caused by leakage from damaged blood vessels, can cause permanent loss of vision

Tip

Long term complications are medical effects that can result after having conditions like diabetes for many years.

Type 2 diabetes has a slightly different cause in that the pancreas stops producing enough insulin, or the insulin that is produced stops working effectively. Type 2 diabetes is usually initially controlled by diet but eventually needs to be controlled by insulin injections or medication (drugs) as the disease progresses.

Type 2 diabetes is usually linked to lifestyle factors such as a poor diet, obesity and lack of exercise.

You need to be able to work out what someone with diabetes should do if their blood glucose level is too high or too low.

Example

What should someone with type 1 diabetes do if their blood sugar rises too high or drops too low?

Answer
If too high – take more insulin (or exercise more to use up more sugar in respiration).

If too low – consume a sugary drink.

Long-term effects and future trends

People who have had diabetes for a long time and whose blood sugar level is not tightly controlled may develop serious long term complications (effects). These include eye damage or even blindness, heart disease and strokes, and kidney damage.

These complications are usually the result of high blood sugar levels damaging the capillaries, which are the fine blood vessels that supply the part of the body involved.

The number of people affected by diabetes is rising rapidly – currently around 10% of the NHS budget is used to treat diabetes and its complications. This is mainly due to the massive increase in people diagnosed with type 2 diabetes in recent years.

This increase is almost entirely due to the increasing number of people who are obese, take little exercise and eat diets high in sugar and fat.

Show you can

Around 10% of the entire NHS budget is used to treat patients with diabetes. Suggest three reasons why this value is so high.

Test yourself

3 What is the scientific name for a chemical messenger that travels in the blood?
4 State **two** ways insulin decreases blood glucose levels.
5 Give **one** difference between type 1 and type 2 diabetes.
6 Give **one** long term complication of diabetes.

Plant hormones

Plants, like animals, respond to changes in the environment and they often do this by using hormones. However, they respond to fewer different types of stimuli and in general the response is slower. Plants respond to the environmental stimuli that have the greatest effect on their growth. Roots grow towards water when a moisture gradient exists. Shoots tend to grow away from the effects of gravity (they grow upwards). Reasons for these responses are fairly obvious as they ensure that plants react in such a way that they receive the best conditions for growth.

The response of a plant shoot to light is called phototropism and this response has been investigated in detail to establish how it occurs.

Phototropism

Most of you will have observed that plants grow in the direction of a light source. Plants left on a windowsill or against the wall of a house usually do not grow straight up, but bend towards the light source, as in Figure 4.9.

This response ensures that the plant stem and leaves receive more light than they otherwise would. This means that more photosynthesis takes place and there will be more growth.

The plant hormone that affects growth and causes phototropism is auxin. This hormone stimulates growth in cells. Figure 4.10 shows that the auxin is produced in the tip of the plant but more passes down the shaded (non-illuminated side) than the illuminated side. This means that the cells in this region get more auxin and therefore grow more (compared to the cells in the side getting most light).

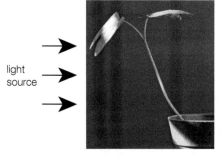

light source

▲ **Figure 4.9** Phototropism

Tip

In normal light conditions (even light on all sides) the auxin is evenly distributed to all sides of the plant stem, therefore the growth of cells is even and the plant grows straight up.

Show you can (?)

Explain how phototropism occurs.

Test yourself

7 Explain the advantage of phototropism to plants.
8 Name the hormone involved in phototropism.

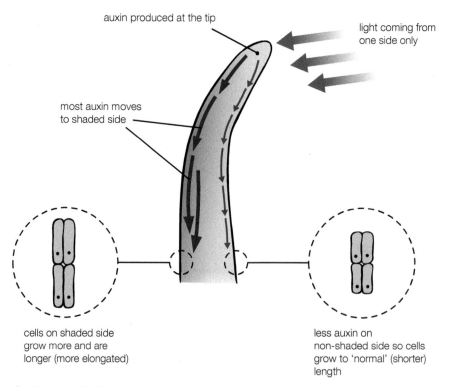

auxin produced at the tip

light coming from one side only

most auxin moves to shaded side

cells on shaded side grow more and are longer (more elongated)

less auxin on non-shaded side so cells grow to 'normal' (shorter) length

▲ **Figure 4.10** The role of auxin in phototropism

1 a) Figure 4.11 shows how nerve impulses pass from receptors to effectors (for example muscles).

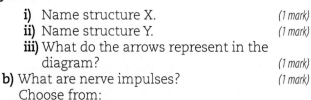

Figure 4.11

    i) Name structure X. *(1 mark)*
    ii) Name structure Y. *(1 mark)*
    iii) What do the arrows represent in the diagram? *(1 mark)*
  b) What are nerve impulses? *(1 mark)*
    Choose from:
    hormones : electrical signals : neurones

2 Figure 4.12 shows a reflex arc.

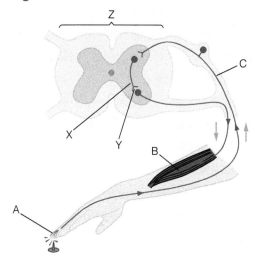

Figure 4.12

  a) Name neurone X. *(1 mark)*
  b) Give the name for the small gap Y. *(1 mark)*
  c) What is structure Z called? *(1 mark)*
  d) Which part of the diagram (A, B or C) is the receptor? *(1 mark)*
  e) Which part (A, B or C) is the effector? *(1 mark)*

3 a) Copy and complete the sentence below about hormones.
    A hormone is a chemical
    ……………………………… that travels in the
    ………………………… to a target organ where it acts.
    (2 marks)
  b) The graph in Figure 4.13 shows how a person's blood glucose level changes over a period of time.

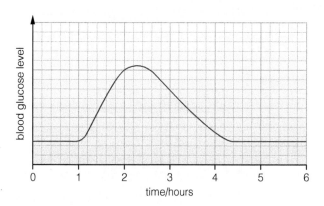

Figure 4.13

    i) Name the organ that produces insulin. *(1 mark)*
    ii) Suggest what caused the blood glucose level to rise after 1 hour. *(1 mark)*
    iii) Describe the role of insulin in causing the blood glucose level to decrease. *(2 marks)*

4 a) Table 4.2 shows the numbers of people being treated for types 1 and 2 diabetes in a hospital over a five year period.

Table 4.2

Year	Number of people treated with diabetes	
	Type 1	Type 2
2005	14	94
2006	15	99
2007	17	105
2008	17	121
2009	19	133
2010	20	141

    i) Calculate the increase in the number of patients with type 2 diabetes between 2005 and 2010. *(1 mark)*
    ii) Calculate the percentage increase in type 2 diabetes between 2005 and 2010. *(1 mark)*
    iii) The table shows that the number of patients with type 2 diabetes is increasing over time. Describe **one** other trend shown by the information in the table. *(1 mark)*
    iv) Suggest **two** reasons for the increase in type 2 diabetes. *(2 marks)*
  b) Give **two** long term complications of diabetes. *(2 marks)*

5 An investigation into how light affects plant seedling growth is summarised in Figure 4.14.

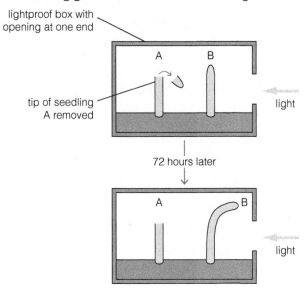

lightproof box with opening at one end

A B

tip of seedling A removed

light

72 hours later

A B

light

Figure 4.14

a) Describe the results shown. *(2 marks)*
b) Name the response shown by seedling B. *(1 mark)*
c) Suggest an explanation for the result for seedling A. *(2 marks)*

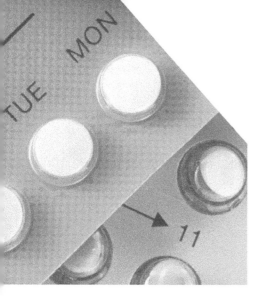

5 The reproductive system

Specification points

This chapter covers sections 1.5.1 to 1.5.5 of the specification. It is about the male and female reproductive systems, the menstrual cycle, pregnancy and contraception.

Reproduction

Living organisms need to be able to reproduce or they would no longer exist. Humans, as almost all animals do, carry out sexual reproduction. Sexual reproduction involves the joining together of two sex cells (gametes) – the egg and the sperm.

The male and female reproductive systems

The male reproductive system

The male reproductive system makes sperm (the male gamete) and is adapted to deliver the sperm into the female reproductive system.

Figure 5.1 shows the male reproductive system and describes the role of each part.

> **Tip**
>
> The scrotum holds the testes outside the body. This means that they are at a slightly lower temperature than body temperature. The sperm grow better at this lower temperature.

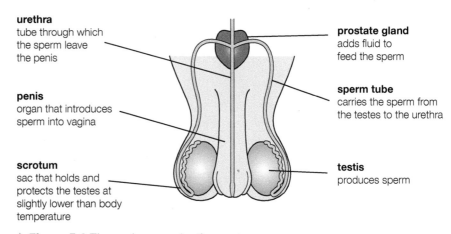

urethra
tube through which the sperm leave the penis

penis
organ that introduces sperm into vagina

scrotum
sac that holds and protects the testes at slightly lower than body temperature

prostate gland
adds fluid to feed the sperm

sperm tube
carries the sperm from the testes to the urethra

testis
produces sperm

▲ **Figure 5.1** The male reproductive system

When a man and a woman have sex, they are in intimate contact and as a consequence the man's penis increases in size and becomes firmer. This enables him to place his penis into the vagina of the woman. During ejaculation, sperm is released by reflex action into the female. Sperm are cells highly adapted for their function. They have a tail that allows the sperm to swim to meet the egg.

The female reproductive system

The female reproductive system is the part of the body that makes and releases eggs. Additionally, if a sperm joins with an egg and pregnancy results, the embryo and foetus are protected and nourished within the female reproductive system until birth.

Figure 5.2 shows the female reproductive system and describes the role of each part.

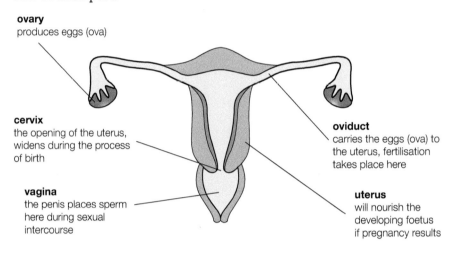

ovary
produces eggs (ova)

cervix
the opening of the uterus, widens during the process of birth

vagina
the penis places sperm here during sexual intercourse

oviduct
carries the eggs (ova) to the uterus, fertilisation takes place here

uterus
will nourish the developing foetus if pregnancy results

▲ **Figure 5.2** The female reproductive system

Following sexual intercourse, the male sperm cell is able to swim out of the vagina, through the cervix, into and through the uterus and into the oviduct where the sperm and egg can fuse (join).

> **Test yourself**
>
> 1 What is the role of the sperm tubes in the male reproductive system?
> 2 Name the part of the female reproductive system that produces eggs.
> 3 In which part of the female reproductive system does fertilisation take place?

Fertilisation and pregnancy

If a sperm and an egg meet and fuse (join) in an oviduct, fertilisation will result.

The fertilised egg becomes the first cell (zygote) of the new individual.

> **Tip**
>
> Although the male can release millions of sperm into the female's vagina during sex, only one sperm is involved in producing a zygote. Most of the sperm do not even make it into the uterus!

The normal number of chromosomes in a cell is described as the diploid number (46 in humans). Sperm and egg cells (nuclei) only contain half the normal number of chromosomes (23 in humans). They are described as being haploid. Fertilisation involves a haploid sperm and haploid egg fusing so that the zygote has the diploid (normal chromosome) number. The scientific name for an egg is ovum.

> **Show you can** ?
>
> Describe the path of a sperm from where it is produced until it leaves the male body.

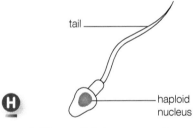

tail

haploid nucleus

▲ **Figure 5.3** A sperm cell

Tip

It is important that all gametes are haploid – if they weren't then every time fertilisation takes place the chromosome numbers in a cell would double!

The single zygote cell formed in fertilisation divides many times and grows into a ball of cells as it travels down the oviduct. This ball of cells becomes attached (implanted) to the wall of the uterus. To enable this to happen the uterus develops a thick lining that holds and nourishes the embryo.

At the point where the young embryo begins to develop in the uterus lining, the placenta and umbilical cord form. A protective membrane, the amnion, forms around the developing embryo. It contains a fluid, the amniotic fluid, within which the growing embryo develops. This fluid cushions the delicate developing embryo. The embryo is referred to as a foetus after a few weeks when it begins to become more recognisable as a baby. Figure 5.5 shows a foetus in the uterus.

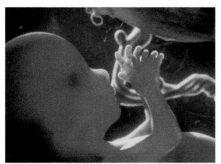

▲ **Figure 5.4** Human foetus at four months, showing the umbilical cord and the placenta

Tip

The placenta is where the exchange of materials takes place between the mother and the foetus.

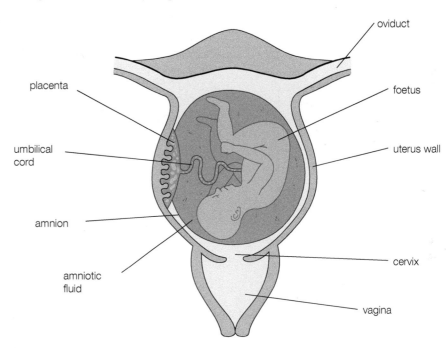

▲ **Figure 5.5** A foetus in the uterus

The foetus cannot breathe when in the amniotic fluid (its lungs will not be developed enough anyway in the early stages), so during pregnancy useful materials including oxygen and dissolved nutrients such as glucose pass from the mother to the foetus through the placenta and umbilical cord. Waste excretory materials including carbon dioxide and urea pass from the foetus back to the mother.

Forty weeks after the ball of cells initially implants, the foetus has developed and grown enough and is ready to be born.

A developing foetus, particularly in the early stages, is very delicate and easily damaged. The uterus wall provides some protection and the amnion and amniotic wall cushion it against knocks.

However, it is important that pregnant women play their part in protecting the developing foetus. This includes being aware and taking account of the fact that scientific evidence shows that consuming alcohol while pregnant can cause harm to the foetus.

The menstrual cycle

The menstrual cycle occurs in females from puberty until the end of reproductive life (usually sometime between the ages of 45 and 55). Each menstrual cycle lasts about 28 days. It is a cyclical event with the release of an ovum, the development of a thick lining on the uterus wall, and the breakdown of this lining (menstruation) occurring in each cycle.

The function of the menstrual cycle is to prepare the reproductive system for pregnancy by controlling the monthly release of an egg and renewing and replacing the uterine lining.

The menstrual cycle is controlled by a number of female hormones. One of the most important hormones is oestrogen. At the start of each menstrual cycle (the onset of bleeding, which we call day one), the level of oestrogen is low. As the cycle progresses the level of oestrogen rises. It peaks in mid cycle, causing the release of an ovum (ovulation).

Another very important hormone is progesterone. The level of progesterone is also low during menstruation and peaks in the days following ovulation. The role of the progesterone is to build up and maintain the thick uterine lining (and the subsequent development of the placenta and other structures associated with pregnancy) should pregnancy occur. Oestrogen is also important in the initial buildup of the uterine lining.

If pregnancy does not occur, the levels of oestrogen and progesterone drop towards the end of the cycle and this causes menstruation to occur. Then the cycle begins again.

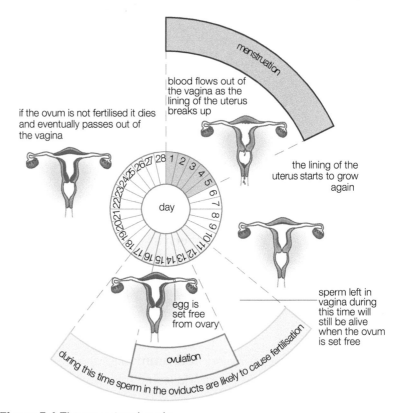

▲ **Figure 5.6** The menstrual cycle

Tip

Oestrogen has two main functions in the menstrual cycle; the initial repair and buildup of the uterus wall and also the stimulation of ovulation. Progesterone is responsible for the buildup and maintenance of the uterine lining, particularly in the period after ovulation and should pregnancy arise.

Tip

Figure 5.6 shows the time periods in the menstrual cycle during which fertilisation of an ovum can occur. You should know that fertilisation can occur if sex takes place in the days immediately before, or immediately after, ovulation.

Test yourself

4 Give two functions of oestrogen in the menstrual cycle.
5 What is the function of progesterone in the cycle?

Show you can

Use information provided in earlier sections to draw a diagram to show how the level of oestrogen changes throughout the menstrual cycle.

Contraception – preventing pregnancy

Many people want to have sex but do not want to have children at that particular time. Pregnancy can be prevented by using contraception.

▲ **Figure 5.7** Contraception: the contraceptive pill and the male condom

Methods of contraception

There are three main types of contraception: mechanical, chemical and surgical. Examples of each and an explanation of how they work, together with their main advantages and disadvantages, are given in Table 5.1.

Table 5.1 Methods of contraception

Type	Example	Method	Advantage	Disadvantage
Mechanical (physical)	male and female condoms	act as a barrier to trap sperm and prevent them swimming up the female's reproductive system	easily obtained and also protect against sexually transmitted infections such as HIV, which can lead to AIDS. Some STIs can lead to infertility if untreated, for example, chlamydia	unreliable if not used properly
Chemical	contraceptive pill	taken regularly by the woman and prevents the ovaries releasing eggs by changing hormone levels	very reliable	can cause some side-effects such as weight gain, mood swings and increased risk of blood clots the woman needs to remember to take the pill daily for around 21 consecutive days in each cycle
	implants	implants are small tubes about 4 cm long that are inserted just under the skin in the arm and release hormones slowly over a long period of time	very reliable can work for up to 3 years	do not protect against STIs can prevent menstruation taking place
Surgical	vasectomy	cutting of sperm tubes, preventing sperm from entering the penis	almost 100% reliable	very difficult or impossible to reverse
	female sterilisation	cutting of oviducts, preventing eggs from moving through the oviduct and being fertilised	almost 100% reliable	very difficult or impossible to reverse

Some people are opposed to contraception but may want to reduce their chances of having children. They can do this by avoiding having sex around the time when the woman releases an ovum each month – this has been called the rhythm or natural method of contraception.

Some people choose this method for religious, ethical or moral reasons but it is much less effective than traditional contraceptive methods. In many women the menstrual cycle is irregular, making it difficult to know exactly when an egg is being released.

You need to be able to explain how each method of contraception works, its main advantage(s) and disadvantage(s).

Example

Explain how female sterilisation works and give one advantage and one disadvantage.

Answer
Both oviducts are cut and this prevents eggs (and sperm coming from the other direction) from passing beyond this point so fertilisation cannot occur. The main advantage is that it is almost 100% reliable. The disadvantage is that it is very difficult or impossible to reverse.

Test yourself

6 Describe how condoms prevent pregnancy.
7 Give **one** advantage of using condoms.
8 Give **one** disadvantage of using the contraceptive pill.

1 Figure 5.8 represents the female reproductive system.

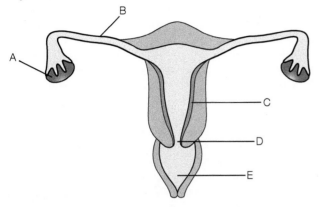

Figure 5.8

  a) Identify parts A and B. *(2 marks)*
  b) Give the letter on the diagram that represents where:
    • fertilisation can take place
    • an embryo (ball of cells) can implant
    • male sperm is deposited during sex.
    (3 marks)

2 a) What is a zygote? *(1 mark)*
  b) Figure 5.9 represents a foetus in the uterus.

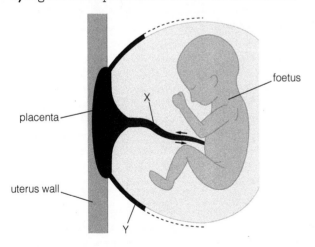

Figure 5.9

  i) Name structures X and Y. *(2 marks)*
  ii) Give **two** substances that pass from the placenta to the foetus. *(2 marks)*
  iii) Describe fully the function of structure Y. *(2 marks)*

3 Figure 5.10 represents the menstrual cycle. **Ⓗ**

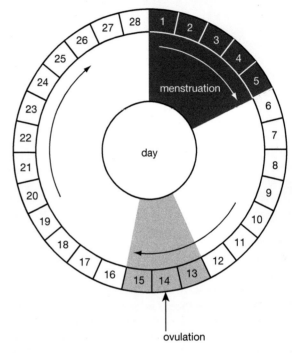

Figure 5.10

  a) Describe **one** change that would occur in the uterus between day 6 and day 12. *(1 mark)*
  b) Name the hormone mainly responsible for this change. *(1 mark)*
  c) Explain fully why it would be possible for a female to get pregnant if sex took place on day 16. *(2 marks)*
  d) Figure 5.11 shows the levels of a hormone during the menstrual cycle.

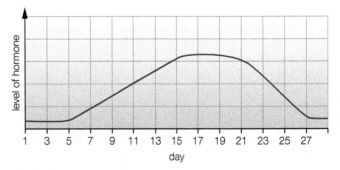

Figure 5.11

  i) Name this hormone. *(1 mark)*
  ii) Give the function of this hormone. *(1 mark)*

4 a) Suggest the effect a blockage in one of the oviducts could have on the chances of a female becoming pregnant. Explain your answer. *(3 marks)*

b) Figure 5.12 represents the male reproductive system.

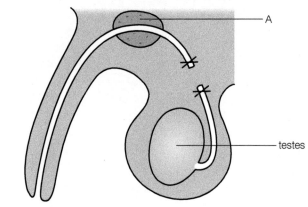

Figure 5.12

i) Name the gland labelled A. *(1 mark)*

ii) Name the method of contraception shown. *(1 mark)*

iii) Using the diagram, explain fully how this method prevents pregnancy. *(2 marks)*

iv) Give **one** disadvantage of this method of contraception. *(1 mark)*

6 Variation and adaptation

Specification points

This chapter covers sections 1.6.1 to 1.6.3 of the specification. It is about types of variation and natural selection.

Variation

If you look at the other members of your class, you will notice that they all look different – this is called variation. There are different types of variation.

Types of variation

Variation can be described as being continuous or discontinuous.

Continuous variation

Continuous variation is a gradual change in a characteristic across a population. This means that there are no clear boundaries between groups (categories) and it may be difficult to decide where one group ends and another starts.

A good example of continuous variation is height in humans (Figures 6.1 and 6.2). There is no clear cut off between being tall or not. Another example is the length of our hand span.

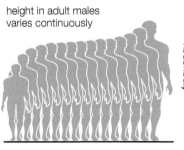

height in adult males varies continuously

▲ **Figure 6.1** Height – an example of continuous variation in humans

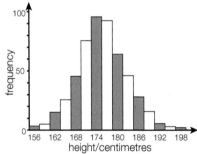

▲ **Figure 6.2** Continuous variation is usually represented in histograms.

Tip

In a histogram, all the categories/groups, for example, different heights or hand span lengths are on the *x*-axis with the bars touching. Number of individuals, such as people/organisms/objects, is on the *y*-axis.

Discontinuous variation

Discontinuous variation occurs when all individuals can be clearly divided into a small number of groups and there are no intermediate states.

Good examples are tongue rolling and hand dominance in humans – everyone either can or cannot roll their tongue, and everyone is either right or left handed.

▲ **Figure 6.3** This man and child can both roll their tongues

Discontinuous variation is usually represented by bar charts.

> **Tip**
>
> In a bar chart, all the categories/groups (such as able or not able to tongue roll or left or right handedness) are on the *x*-axis with equal sized gaps between the bars. Number of individuals, for example people/organisms/objects, is on the *y*-axis.

Causes of variation

Variation can be genetic (due to our genes) or environmental (due to environmental factors).

Some characteristics or features are purely genetic, such as human blood groups – we can only be A, B, AB or O. However, if we grow cuttings from the same geranium plant in different environmental conditions (such as different amounts of light), then any differences in the plants must be environmental as they are all genetically the same.

Many features have a combination of genetic and environmental causes. While our height is largely genetically controlled, the actual height we reach depends on environmental factors such as our diet and how much we exercise.

> **Tip**
>
> If you can clearly allocate an individual to a particular group (without debate) then it is discontinuous variation.

▲ **Figure 6.4** Tongue rolling – an example of discontinuous variation in humans

> **Tip**
>
> You should be able to explain why shoe size is discontinuous yet foot length is continuous variation.

Show you can

Draw a table with two columns. Use continuous variation and discontinuous variation as the two headings. Place the following examples of variation in the correct column: blood group, hand span, shoe size, weight, height.

Test yourself

1 What is continuous variation?
2 What is discontinuous variation?
3 Give the two causes of variation.

Natural selection and evolution

Natural selection

In nature, adaptations in living organisms are essential for survival and success in all habitats. Students should be able to work out the adaptations of organisms from information given.

Example

Using the photograph, give two adaptations to the environment in polar bears.

▲ **Figure 6.5** Polar bear and cub

Answer
They have thick fur to reduce heat loss and their white fur gives them camouflage and allows them to blend in with the snow.

Tip

Individuals (phenotypes) in a population vary; variation is needed for natural selection. If every organism was exactly the same, then none would be better adapted than the rest!

Adaptations in organisms are even more important when they compete with each other for resources.

Competition ensures that the best adapted individuals will survive. For example, the larger seedlings growing in a clump of plants will be able to obtain resources such as light, nutrients and water more easily than the smaller seedlings. As a result of this competition, the stronger individuals will survive and the weaker ones may die out.

This competition for survival, with the result that the better equipped individuals survive, summarises Charles Darwin's theory of natural selection.

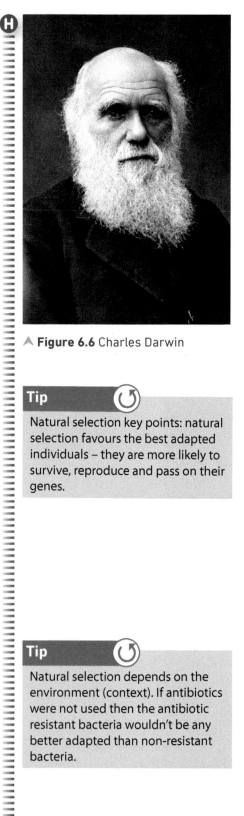

▲ **Figure 6.6** Charles Darwin

Tip

Natural selection key points: natural selection favours the best adapted individuals – they are more likely to survive, reproduce and pass on their genes.

Tip

Natural selection depends on the environment (context). If antibiotics were not used then the antibiotic resistant bacteria wouldn't be any better adapted than non-resistant bacteria.

Charles Darwin and the theory of natural selection

Charles Darwin devoted much of his life to scientific research. As part of his research, he spent five years on the *HMS Beagle* as it travelled to South America. Darwin was greatly influenced by the variety of life he observed on his travels and, in particular, by the animals of the Galapagos Islands. Darwin's famous account of natural selection, *On the Origin of Species*, was published in 1859.

Darwin's main conclusions about natural selection can be summarised as:

▶ There is variation among the phenotypes (individuals) in a population.

▶ If there is competition for resources there will be a struggle for existence.

▶ The better-adapted phenotypes survive this struggle or competition. This leads to survival of the fittest and these (fittest) individuals are more likely to reproduce and pass genes on to the next generation.

It is useful to look at an example of natural selection in action to highlight the key features of Darwin's theory.

Antibiotic resistance in bacteria

When bacteria are treated with an antibiotic such as penicillin, most of them are killed. However, a small number (the best adapted phenotypes) may survive, probably because they have a gene (caused by a mutation) that provides resistance. Very soon the resistant bacteria are the only ones remaining, as they are the only ones surviving and passing their beneficial genes on to their offspring. Figure 6.7 shows the development of antibiotic resistance in bacteria.

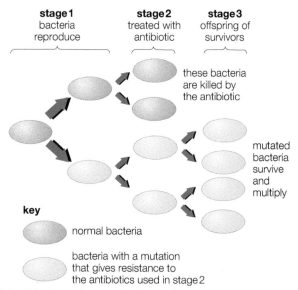

▲ **Figure 6.7** Antibiotic resistance in bacteria

The link between natural selection and evolution

Darwin used the theory of natural selection to explain the process of evolution. He suggested that species have changed gradually through time in response to changes in the environment and that evolution is a continuing process. It is natural selection that causes these changes over time.

Evolution can lead to changes within one species over time or can lead to the formation of new species.

Fossils

There is so much evidence for evolution now that most scientists accept that evolution has taken place and that it is still taking place. One of the best types of evidence is fossil evidence. Fossils are the remains of living organisms that have been preserved (usually in rocks) for millions of years.

The fossil of an *Archaeopteryx* in Figure 6.8 is one of the most important fossils ever found. This and other similar fossils showed how birds evolved from reptiles (dinosaurs).

Therefore, fossils not only show what a particular animal or plant looked like millions of years ago. They often show how the organism (species) changed over a long period of time.

For example, the fossil record of horses is extensive and, because of this, it shows how horses evolved over time in a changing habitat (see Table 6.1). The horses' fossil record shows the various stages, not just the 'before' and 'after'.

Tip

Evolution can be defined as change in a species over time.

▲ **Figure 6.8** Fossil of *Archaeopteryx*

Tip

As it is possible to date rocks back to when they were formed (and when the organism was fossilised) it is possible to fairly accurately date the age of the fossil.

Table 6.1 Evolution in horses

Time	Habitat	Hoof size	Size of horse	Explanation
↓	wet marshes ↓ dry grasslands	large ↓ small	small ↓ large	As the marshes dried to become grassland, smaller hooves were advantageous as they allowed horses to run faster and escape predators. Similarly, the extra height as horses grew larger meant they could spot predators more easily.

Tip

It is important to understand that the phenotype of the horse changed in many very small steps over a very long period of time – as is typical of evolutionary change.

Tip

Natural selection acted on hoof size and animal size, favouring the adaptation that suited the habitat best at any particular time. As the habitat changed, the best adapted phenotype did too.

Extinction

Species are extinct if there are no living examples left. Many species have become extinct and often we only know they did exist in the past due to the discovery of fossils. Examples of extinct species include the dodo, dinosaurs and the woolly mammoth.

Tip

Species can become extinct if they fail to adapt to environmental change or their 'surroundings'.

▲ **Figure 6.9** Woolly mammoths

Species that are not quite extinct but are at risk are called endangered species. For example, a number of big cat species (such as tigers and leopards) are endangered.

Species can become extinct for many reasons. These include:

▶ climate change or natural disasters – climate change today is causing plants and animals to become extinct

▶ hunting by humans, for example, the dodo was hunted until it became extinct

▶ hunting by animals introduced by humans to areas where they are not normally found

▶ spread of diseases

▶ loss of habitat. This is causing a lot of species to become extinct today. The loss of habitat is often caused by human activities, such as deforestation and land-clearing for towns and cities.

Tip

The list shows that humans are responsible for many of the extinctions that take place today – in the context of most extinctions, the activities of humans bring about 'environmental change'.

Example

It is thought that a huge asteroid up to 10 km wide striking the Earth in the Gulf of Mexico caused the extinction of the dinosaurs and around 75% of other plant and animal species.

As a result of the asteroid's impact so much dust and debris ended up in the atmosphere that the Sun's light and heat couldn't penetrate the atmosphere to reach the Earth. Therefore photosynthesis couldn't take place and many plants and animals including the dinosaurs died out.

Show you can

Using the example of antibiotic resistance in bacteria, explain the difference between natural selection and evolution.

Test yourself

4 Describe what is meant by antibiotic resistance.
5 Explain what is meant by the term evolution.
6 State **two** ways in which humans have caused extinctions of other species.

1 Figure 6.10 shows some Friesian cattle grazing in a field.

Figure 6.10

a) Name the type of variation shown by coat colour in the cattle. *(1 mark)*
   Choose from:
   discontinuous   continuous   environmental
b) Define this type of variation. *(1 mark)*
c) Suggest **one** other example of variation shown by the cattle in the field. *(1 mark)*

2 a) Figure 6.11 shows how leaf width varies in a particular type of plant.

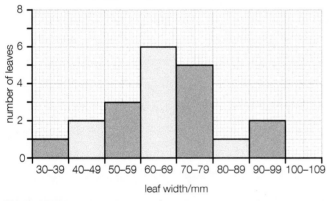

Figure 6.11

i) What is the most common leaf width in these plant leaves? *(1 mark)*
ii) How many leaves were sampled in total? *(1 mark)*
iii) Name the type of variation shown by leaf width. *(1 mark)*

b) In a particular class of Year 11 students, twelve could roll their tongue and eight could not.
   i) What percentage of students could roll their tongue? *(2 marks)*
   ii) Why is tongue rolling described as discontinuous variation? *(1 mark)*
   iii) Name the type of graph used to show discontinuous variation. *(1 mark)*

3 Scientists cultured two types of bacteria (A and B) in a beaker. Figure 6.12 shows how the numbers of the two types changed after an antibiotic was added to the beaker.

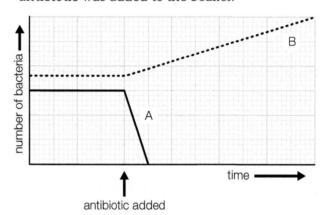

Figure 6.12

a) Describe fully the changes in numbers of bacteria A and B after the antibiotic was added. *(2 marks)*
b) Explain the change in number of B. *(3 marks)*
c) Name the process that this investigation demonstrates. *(1 mark)*

4 Most fossils are found in rock. For example, the *Archaeopteryx* fossils described earlier in this chapter were found in quarries in Germany.
a) Suggest why many fossils are found in quarries or where new roads are being built. *(1 mark)*
b) Describe fully how fossils provide evidence for evolution. *(3 marks)*

5 a) Table 6.2 below shows the number of extinctions that have occurred in a country over the last 100 years.

Table 6.2 Number of extinctons

Year	Number of extinctions
1920	1
1940	2
1960	15
1980	22
2000	46

i) Copy the graph grid (Figure 6.13) and complete a line graph of this information. *(3 marks)*

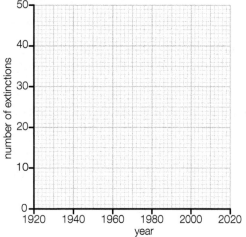

Figure 6.13

ii) In which twenty-year period was there the largest increase in extinctions? *(1 mark)*

iii) Predict how many extinctions there will be by the year 2020. *(1 mark)*

b) The dodo was a large flightless bird that lived on the island of Mauritius in the Indian Ocean. When sailors visited the island during the 1500s and 1600s they hunted the dodo and removed much of its forest habitat. Additionally they brought pigs and rats to the island, both of which fed on the dodo's eggs. By the end of the 1600s the dodo had become extinct.

i) Give **three** ways in which humans contributed to the extinction of the dodo. *(3 marks)*

ii) Suggest **one** reason why the dodos were an easy target for human hunters. *(1 mark)*

7 Disease and body defences

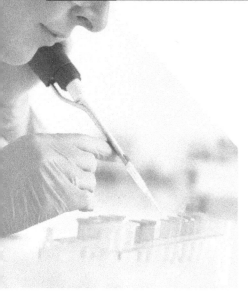

Specification points

This chapter covers sections 1.7.1 to 1.7.7 of the specification. It is about types of microorganisms, the body's defence mechanisms, antibiotics, antibiotic-resistant bacteria, the development of medicines, and alcohol and tobacco.

Types of microorganisms

A communicable disease is a disease that can be passed from one organism (person) to another.

Tip

Communicable diseases are also described as infectious diseases.

Most communicable diseases are caused by microorganisms such as bacteria, viruses and fungi. Harmful microorganisms are called pathogens.

Table 7.1 shows some diseases caused by microorganisms and how they can be spread (what makes them communicable). The table below also shows how they can be prevented or treated.

Table 7.1 Communicable (infectious) diseases – their cause, spread and prevention or treatment

Microbe	Type	Spread	Control/prevention/treatment
HIV (which can lead to AIDS)	virus	exchange of body fluids during sex infected blood	using a condom will reduce risk of infection, as will drug addicts not sharing needles currently controlled by drugs
Colds and flu	virus	airborne (droplet infection)	flu vaccination for targeted groups
Human papilloma virus (HPV)	virus	sexual contact	HPV vaccination given to 12–13 year olds to protect against developing cervical cancer
Salmonella food poisoning	bacterium	from contaminated food	always cooking food thoroughly and not mixing cooked and uncooked foods can control spread treatment by antibiotics
Tuberculosis	bacterium	airborne (droplet infection)	BCG vaccination if contracted, treated with drugs including antibiotics
Chlamydia	bacterium	sexual contact	using a condom will reduce risk of infection treatment by antibiotics
Athlete's foot	fungus	contact	reduce infection risk by avoiding direct contact in areas where spores are likely to be present, e.g. wear 'flip flops' in changing rooms/swimming pools
Potato blight	fungus	spores spread in the air from plant to plant, particularly in humid and warm conditions	crop rotation and spraying plants with a fungicide

Figure 7.1 shows the spray of moisture and particles spread through the air when a man sneezes. If he has a cold or flu, this is how 'droplet' infection can occur.

Figure 7.1 shows the spray of moisture and particles spread through the air when a man sneezes. If he has a cold or flu, this is how 'droplet' infection can occur.

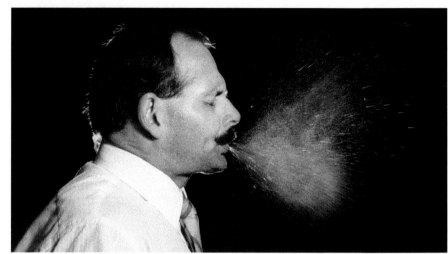

▲ **Figure 7.1** A man sneezing and possibly spreading microorganisms which cause disease

The body's defence mechanisms

There are microorganisms all around us, yet the majority of people are not sick most of the time. This is because our bodies have good defences against disease. We can both stop harmful microorganisms gaining entry to our body and also destroy them if they get into the blood stream.

Stopping microorganisms gaining entry

We have a number of methods to stop harmful pathogens getting into the body. These methods include:

▶ the skin – the skin provides a covering to stop microorganisms getting in

▶ mucous membranes – the lining of the nose and other parts of the respiratory system have a fine lining of mucus that traps microorganisms

▶ blood clotting – stops blood escaping and prevents microorganisms getting in through cuts.

Antibodies

Antibodies are produced by special white blood cells called lymphocytes to help us defend against microorganisms in the blood. Microorganisms have special marker chemicals called antigens that 'alert' lymphocytes and cause the body to produce the right antibodies. Figure 7.2 shows a very important feature of antibodies – they have to fit exactly with the antigens of the microbe. This is why we have different antibodies for different diseases.

As the antigens on a particular microorganism and the antibodies used to combat that microorganism are complementary in shape it is possible to work out the shape of one from the other. See the examples in Figure 7.3.

Tip

Potato blight is a plant disease that affects the potato and similar plants – all the other communicable diseases in Table 7.1 are passed between humans.

the **lymphocytes** produce **antibodies** in response to the antigen – the antibodies produced are **complementary** in shape to the shape of the antigen – they fit together like a lock and key

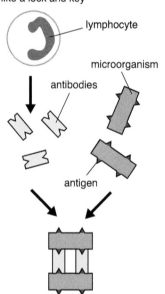

the antibodies 'latch on' to the antigens of the microorganism causing them to clump together – the immobilised microorganism can then be destroyed by other white blood cells called **phagocytes**

▲ **Figure 7.2** How antibodies work

Tip

Antigens are 'marker' chemicals on the surface of microorganisms that trigger the production of antibodies.

Example ⬅

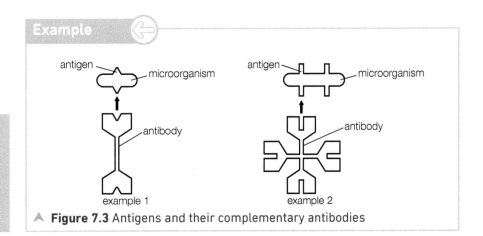

▲ **Figure 7.3** Antigens and their complementary antibodies

Once the microbes are clumped together they are destroyed by the phagocytes (Figure 7.4), which are the other main type of white blood cell. This process is called phagocytosis.

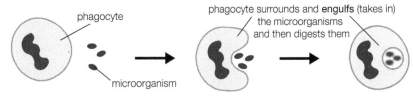

▲ **Figure 7.4** Phagocytosis

The antibody/antigen reaction described in the section above is a typical response to being infected by a bacterium or a virus. The infected individual is often ill for a few days before the antibody numbers are high enough to provide immunity. This is known as the primary response.

However, once infected, the body is able to produce memory lymphocytes that remain in the body for many years. This means that if infection by the same type of microorganism occurs again, the memory lymphocytes will be able to produce antibodies very fast to stop the individual catching the same disease again. This very rapid response when infected by a pathogen a second time is called the secondary response. Figure 7.5 shows the level of antibodies created during the primary and secondary responses.

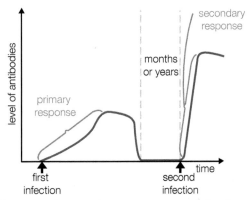

▲ **Figure 7.5** Primary and secondary responses to infection

Immunity

Individuals who are protected against a particular infection or disease are described as being immune to that disease. Most people will be immune to a number of diseases. If someone is immune this means that his or her antibody levels are high enough (or high enough levels can be produced quickly enough) to combat the microorganism should it gain entry to the body again.

There are two types of immunity.

▶ **Active immunity** is where the body produces the antibodies used to combat the infectious microorganism. This type of immunity is slower acting but usually lasts for a very long time (see Figure 7.6).

▶ **Passive immunity** is when antibodies from another source (for example produced by pharmaceutical companies) are injected into the body. These are fast acting but only last for a short period of time (see Figure 7.7).

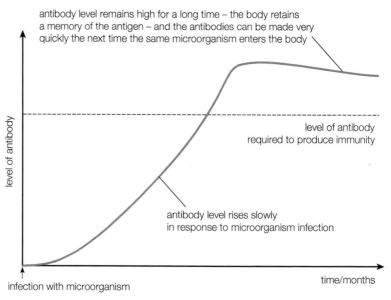

Figure 7.6 Active immunity as a result of having the disease

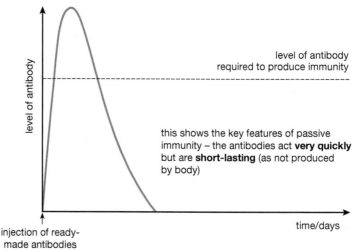

Figure 7.7 Passive immunity by the injection of ready-made antibodies

Test yourself

1 What is a communicable disease?
2 Name two types of human disease spread by droplet infection.
3 Name the two types of white blood cell that help defend against disease.
4 Give two features of passive immunity.

Show you can

The antibodies we produce to combat flu are different to the antibodies we produce for the cold. Explain why.

Antibiotics

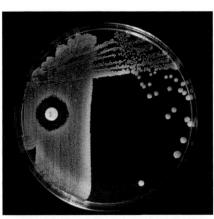

▲ **Figure 7.8** Agar plate showing a clear area (around the white circle on the left) where an antibiotic has killed bacteria growing on the agar

Antibiotics, for example penicillin, are chemicals produced by fungi that are used against bacterial diseases to kill bacteria or reduce their growth.

Most people have had antibiotics at some time in their lives to defend against bacterial conditions such as septic throats or infected wounds in the skin. The effect of an antibiotic on bacteria can be seen in Figure 7.8.

Antibiotics are not as specific as antibodies in that they are normally not designed to combat only one type of bacteria – they usually act against a range of bacteria and they act in a different way to antibodies. However, different types of antibiotics have different effects against different bacteria. For this reason a GP may prescribe different antibiotics at different times for the same patient if the bacterial infections are different.

> **Tip**
>
> Antibiotics cause harm to bacteria. They have no effect on virus diseases, for example colds and flu.

Antibiotic-resistant bacteria

Sometimes bacteria can mutate and this makes them resistant to antibiotics. It is important to note that the mutations occur randomly – they are not caused by the antibiotics. However, in the presence of antibiotics, 'normal' bacteria are killed leaving only the resistant mutated bacteria, which then reproduce and spread. Superbugs such as methicillin-resistant *Staphylococcus aureus*, or MRSA, are so called because they are resistant to many antibiotics. Superbugs can be a big problem in hospitals because antibiotics will not kill them.

> **Tip**
>
> Mutations in the DNA of bacteria give them new properties that make them resistant to antibiotics.

Overuse of antibiotics has led to a big increase in the number of antibiotic-resistant bacteria. In the past, antibiotic-resistant bacteria have been a particular problem in hospitals for a number of reasons including:

▶ many antibiotics are used in hospitals

▶ in hospitals, people are already ill and may have weak immune systems

▶ patients may have open wounds that are easily infected

▶ microbes can easily spread from patient to patient.

What can hospitals do to stop MRSA spreading?

They can:

▶ apply very strict hygiene conditions such as washing hands regularly and mopping up spills of blood or body fluids immediately

▶ isolate patients who have a 'superbug', keeping them away from other patients

▶ not overuse antibiotics.

> **Tip**
>
> Overuse of antibiotics includes using antibiotics on virus diseases (where they won't work) and also using antibiotics on 'minor ailments' where the patient is capable of recovering quickly without an antibiotic.

Development of medicines

Medicines are drugs that are used to make us better or reduce pain. Antibiotics such as penicillin are only one type of a wide range of medicines that are available today.

Penicillin

Penicillin was the first antibiotic to be developed. In 1928, **Alexander Fleming** was growing bacteria on agar plates and he noticed that one of his plates had become contaminated with mould (fungi). This is common when culturing bacteria unless great care is taken to avoid contamination.

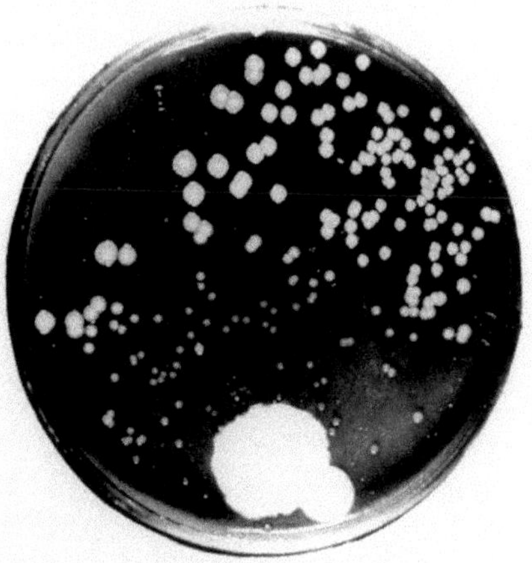

The growth of mould did not surprise Fleming, but he was surprised when he noticed that the bacteria he was culturing did not grow well close to the edges of the mould, as shown in Figure 7.9 and Figure 7.10. He concluded that the mould produced a substance that prevented the growth of the bacteria. As the fungus causing the contamination was called *Penicillium*, the antibacterial substance was called penicillin and the first antibiotic was developed.

▲ **Figure 7.9** This photograph shows the original Petri dish in which Fleming noticed the effect fungal contamination had on bacterial growth. This photograph shows the importance of careful observation in science!

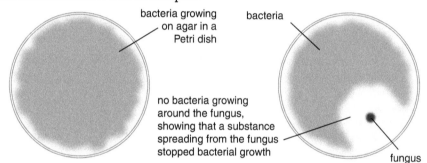

bacteria growing on agar in a Petri dish

bacteria

no bacteria growing around the fungus, showing that a substance spreading from the fungus stopped bacterial growth

fungus

▲ **Figure 7.10** Fleming's discovery

Fleming carried out some work with his antibacterial substance on animals, but his progress was hindered because he was unable to produce a pure form of the substance. In the early 1940s, two other scientists, Florey and Chain, were able to isolate a pure form of penicillin and its large scale production began. Penicillin has been in use since then but is now only one of a large number of antibiotics in use.

▲ **Figure 7.11** Howard Florey

▲ **Figure 7.12** Ernst Chain

Making new medicines and drugs

We have already seen the role Alexander Fleming played in the discovery of penicillin. However, before a medicine or drug can be made available to the public there are a number of stages that must take place. These stages usually take a very long time (often many years) and cost a very large amount of money.

Some of the stages are described below.

In vitro testing

Tip

In vitro testing is really to check that the drug works on living cells without causing them too much harm.

This is the testing of a very early version of the drug on living cells in the laboratory. Unless the scientists find that the drug works in living cells and that the cells are not harmed, further testing is unlikely to happen. *In vitro* testing can be very expensive as it is very much a 'trial and error' process. Highly trained scientists and usually a lot of very expensive equipment are needed.

Animal testing

Animal testing is usually the next stage. It is an important process as it tests the drug on whole animals with complete immune systems. Animal testing is usually carried out on animals that have similar body systems to humans (other mammals). Animal testing means scientists do not need to test on humans at this early stage of the drug's development since this could be dangerous to health.

However, many people are opposed to animal testing. They argue that it is cruel to test on animals and that the animals used are a different species to humans, so proving that a drug is safe in animals is no guarantee that it is safe to use on humans. Although this is a complex ethical issue, it appears that animal testing will continue until some suitable alternative is found.

▲ **Figure 7.13** *In vitro* testing in the laboratory

Clinical trials

Clinical trials follow animal testing. In clinical trials, the drug is tested on human volunteers for three main reasons:

▶ to see if the drug actually works on humans

▶ to see if there are any side effects

▶ to check the dosage that should be used.

Tip

As humans are a different species to the animals that a drug is tested on, there is no guarantee that a drug will work on humans until it is actually tested on them.

Volunteers wanted for medical research

Volunteers should be between 20 and 30 years old and currently taking no medication.

Volunteers will be asked to spend one night in a clinic and take part in a number of follow-up visits.

For further details contact …

▲ **Figure 7.14** Advert asking for clinical trial volunteers

Show you can

Explain why the process of drug development is very expensive.

Initially the drug is tested on a very small number of people but in due course much larger numbers are involved. The trials usually involve patients who have the condition the drug is targeting and volunteers (who are often paid a fee). Figure 7.14 shows an advert asking for volunteers to take part in a clinical trial. Would you be willing to act as a volunteer in a drug trial?

Following clinical testing, if it is clear that the drug works in the way it was intended and that the patients do not suffer serious side effects, the drug may be licensed for use. Some drugs may be licensed even though they do have side effects. This could happen if the benefits of the drug outweigh the harm it causes.

Test yourself

5 What is an antibiotic?
6 Name the first antibiotic discovered.
7 Describe what *in vitro* testing is.
8 Name the stage that follows *in vitro* testing in drug development.

Alcohol and tobacco

Alcohol and tobacco are also 'drugs' but in these cases they can actually cause harm to the body.

Misuse of drugs – alcohol

Many people drink alcohol in moderation and are unlikely to suffer serious harm. However, other people, including many teenagers, drink too much alcohol and can cause harm to themselves and others.

Long term excessive drinking of alcohol can damage the liver as well as many other parts of the body. Drinking heavily during pregnancy can cause serious damage to the foetus including brain damage (foetal alcohol syndrome).

Binge drinking is a particular problem. This occurs when a large amount of alcohol is taken over a short period of time, for example, on one night out (Figure 7.15).

▲ **Figure 7.15** A consequence of binge drinking

Misuse of drugs – tobacco smoke

Smoking can seriously damage health, as summarised in Table 7.2.

Table 7.2 The effects of tobacco smoke on the body

Substance in cigarette smoke	Harmful effect
Tar	causes bronchitis (narrowing of the airways into the lungs), emphysema (damage to the gas exchange surfaces in the lungs) and lung cancer
Nicotine	addictive and affects heart rate
Carbon monoxide	combines with red blood cells to reduce the oxygen carrying capacity of the blood

Tip

Bronchitis can leave affected individuals struggling for oxygen as not enough oxygen may reach the gas exchange surfaces due to the airways (bronchi and bronchioles) being constricted.

Tip

Emphysema damages the gas exchange surfaces in the lungs. Gas exchange surfaces are the walls of the alveoli (air sacs) where oxygen passes from the lungs into the blood. If the walls of the alveoli are damaged, less oxygen can pass into the blood.

Tip

Carbon monoxide combines with red blood cells, reducing the number of red blood cells available to carry oxygen. This can result in a shortage of oxygen reaching the body cells and therefore there will be less oxygen available for respiration.

The introduction of smoking bans in many countries has proved very effective. It both encourages smokers to stop and significantly reduces people being affected by passive smoking. The use of E-cigarettes is enabling many people to stop smoking tobacco. However, not everyone is in favour of E-cigarettes as many contain nicotine and some people suggest that they encourage people to take up smoking tobacco.

Test yourself

9 What is binge drinking?
10 Give **one** harmful effect of drinking too much alcohol.
11 Name **three** diseases caused by tobacco smoke.

Show you can

Explain how the different effects of tobacco smoke can cause there to be a shortage of oxygen in the blood.

Practice questions

1 a) Copy and complete Table 7.3 about communicable diseases and how they are spread. *(4 marks)*

Table 7.3

Disease	Type of microorganism	Method of spread
salmonella		in contaminated food
HPV		sexual contact
potato blight		

b) Describe how tuberculosis can be:
- prevented *(1 mark)*
- treated. *(1 mark)*

2 The flu is a communicable virus disease that can make people ill for a number of days. However, most people recover from the flu.
a) What is meant by the term communicable disease? *(1 mark)*
b) i) Explain why it takes a number of days for people to recover from the flu. *(1 mark)*
ii) Describe the role of antibodies in helping people combat the flu. *(3 marks)*
iii) Give **two** differences between active and passive immunity. *(2 marks)*

3 a) Copy and complete Figure 7.16 to show the effect of an antibiotic on this group of bacteria. *(1 mark)*

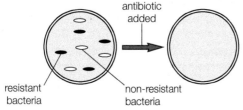

Figure 7.16

b) i) Explain what is meant by the term 'superbug'. *(1 mark)*
ii) State the main reason for the increase in the number of 'superbugs'. *(1 mark)*
iii) Suggest **two** reasons why 'superbugs' are a particular problem in hospitals. *(2 marks)*
iv) Give **two** ways in which the incidence of 'superbugs' in hospitals can be reduced. *(2 marks)*

4 a) Animal testing, clinical trials and *in vitro* testing are stages in the development of drugs and medicines.
i) Put these stages in the correct order when testing drugs. *(1 mark)*

ii) Clinical trials test the drug on human volunteers. Give **two** reasons why it is necessary to test on humans. *(2 marks)*
b) Table 7.4 shows the number of alcohol units three young men drank over a six-day period.

Table 7.4

	Ben	Liam	Sean
Monday	0	0	1
Tuesday	0	0	2
Wednesday	1	0	1
Thursday	1	2	1
Friday	2	6	2
Saturday	1	0	2

i) Which young man drank the most alcohol over this period? *(1 mark)*
ii) Which young man was binge drinking during this period? Explain your answer. *(2 marks)*
iii) Name a body organ that can be harmed by drinking too much alcohol. *(1 mark)*

5 a) Graphs 1 and 2 in Figure 7.17 show how the number of people smoking and the number of people diagnosed with lung cancer have changed over a 50-year period in a large city.

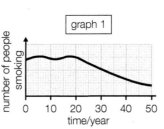
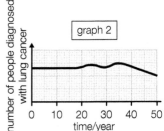

Figure 7.17

i) What is the evidence from the graphs to suggest that smoking is a cause of lung cancer? *(1 mark)*
ii) What is the evidence that suggests that it takes a number of years for lung cancer to develop? *(1 mark)*
b) i) Name the component in tobacco smoke that causes lung cancer. *(1 mark)*
ii) Name **one** other harmful effect that this component causes. *(1 mark)*

Specification points

This chapter covers sections 1.8.1 to 1.8.9 of the specification. It is about photosynthesis, the role of the Sun as an energy source, food chains and food webs, monitoring environmental change, competition and human activity on Earth.

Photosynthesis

Photosynthesis is a chemical reaction that takes place in the chloroplasts of plants. Chloroplasts contain chlorophyll, which is the chemical that absorbs light energy.

Photosynthesis needs light energy from the Sun. Photosynthesis uses water and carbon dioxide to make glucose with oxygen given off as a waste product. The word equation for photosynthesis is:

carbon dioxide + water → glucose + oxygen

The balanced symbol equation is:

$$6CO_2 + 6H_2O \rightarrow C_6H_{12}O_6 + 6O_2$$

As photosynthesis requires (light) energy to work it is an endothermic reaction.

The starch test

One way of showing that photosynthesis is taking place is to show that starch is being produced. The glucose that is produced during photosynthesis is usually converted to starch in the leaf for short term storage.

You can show that photosynthesis is taking place by carrying out the starch test on a leaf, as shown in Figure 8.1 and described in Table 8.1.

Table 8.1 The starch test

Step	What happens
1 Remove a leaf from a plant and place it in boiling water.	This will kill the leaf and stop any further reactions.
2 Boil the leaf in ethanol (alcohol). Do this in a water bath using very hot water from an electrical kettle. You should not use a Bunsen in this investigation as the ethanol is flammable.	This removes the green chlorophyll from the leaf.
3 Dip the leaf in boiling water again.	This makes the leaf soft and less brittle.
4 Spread the leaf on a white tile and add iodine solution.	If starch is present the iodine will turn from yellow-brown to blue-black.

Tip

Endothermic reactions are reactions that require energy to be absorbed (taken in) to work.

Tip

Photosynthesis takes place mainly in the leaves of plants. This is because it is in the leaves that most of the chloroplasts (and therefore chlorophyll) are found.

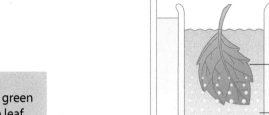

Tip

It is important to remove the green colouring (chlorophyll) in the leaf as it would mask the colour change when iodine is added.

▲ **Figure 8.1** Preparing a leaf for the starch test

Prescribed practical

Prescribed practical B3: Investigate the need for light and chlorophyll in photosynthesis by testing a leaf for starch.

Showing that light is necessary for photosynthesis

Procedure

1 Before carrying out experiments on photosynthesis it is important to destarch the plant. This means removing any starch that is already there before the experiment. To do this, place the plant in the dark for two days. Destarching the plant is done to ensure that any starch produced is only produced during the investigation.

2 A leaf is then partially covered with foil to allow some parts of the leaf to access light while other parts are kept in the dark.

3 Leave the plant in bright light for at least 6 hours.

4 Test the leaf for starch using the starch test.

Typical results and questions

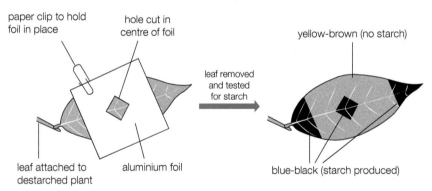

▲ **Figure 8.2** Experiment to show that light is required for photosynthesis to occur

1 State the main conclusions that can be drawn from these results.

2 Why was it necessary for the plant to be left in bright light for at least 6 hours?

Showing that chlorophyll is necessary for photosynthesis

Some plants have leaves that are part green and part white as shown in Figure 8.3. These leaves are described as variegated leaves.

Procedure

1 Destarch a variegated plant.

2 Leave in bright light for at least 6 hours.

3 Carry out the starch test on a variegated leaf.

Questions

1 Describe and explain the results you would expect to get.

2 Suggest why very few plants have variegated leaves in nature.

It is possible to combine the two parts of this investigation. This would involve covering part of a variegated leaf with foil.

▲ **Figure 8.3** Variegated leaves

Showing that oxygen is produced

Using apparatus similar to Figure 8.4 it is possible to demonstrate that oxygen is produced in photosynthesis.

The rate of photosynthesis will affect the rate at which the bubbles of oxygen are given off and this can be used to compare photosynthesis rates in different conditions. For example, by moving the position of the lamp it is possible to investigate the effect of light intensity on photosynthesis.

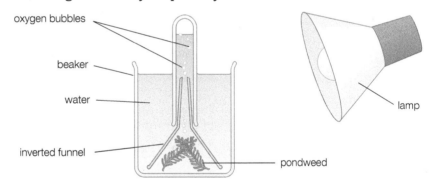

▲ **Figure 8.4** Investigation to show that oxygen is produced during photosynthesis

Test yourself

1 Give the word equation for photosynthesis.
2 Why is a leaf boiled in ethanol during the starch test?
3 Give the colour change for iodine if starch is present.

Show you can ❓

Describe how you could compare the amount of oxygen given off by pondweed in different intensities of light.

Food chains and food webs

Food chains

Figure 8.5 shows a sequence of living organisms through which energy passes. It is an example of a food chain. Food chains show the feeding relationships and energy transfers between a number of organisms. The diagram also shows that the Sun is the primary (original) source of energy.

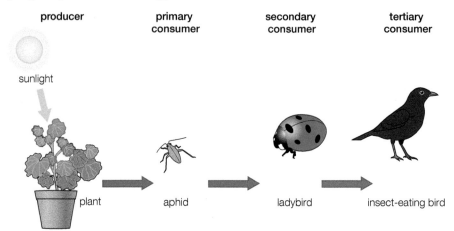

| producer | primary consumer | secondary consumer | tertiary consumer |

plant → aphid → ladybird → insect-eating bird

▲ **Figure 8.5** Energy and feeding relationships

This example shows that in a food chain the first organism is the producer. The producer captures light energy from the Sun and makes this available for other organisms by the process of photosynthesis. It does this by providing the food (energy) for the primary consumer (and eventually the other animals in the food chain).

The primary consumer is the animal that feeds on producers. A secondary consumer then feeds on primary consumers, and so on.

Food chains are very simplistic in that they do not show the range of different feeding relationships that usually exist. For example, very few animals have only one food source.

> **Tip** ↻
>
> Every food chain must start with a producer. The producers are the only organisms that can make energy from sunlight by photosynthesis.

> **Tip** ↻
>
> Consumers are animals that feed on other living things.

Food webs

Food webs show how a number of food chains are interlinked. Figure 8.6 shows a grassland food web.

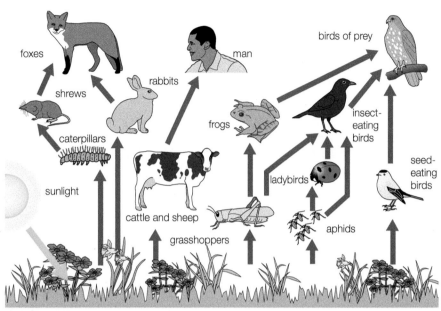

▲ **Figure 8.6** A grassland food web

The arrows in food chains and food webs show the transfer of energy through ecosystems. The arrows also show what eats what – the head of the arrow shows the animal that eats the organism at the start of the arrow. (An arrow from the Sun to a producer is an exception – it still shows the direction of energy transfer but the plant obviously doesn't eat the Sun!)

You need to be able to identify the producer and consumers in food chains and food webs.

Example ⬅

Slugs feed on green plants, and birds such as the thrush feed on the slugs. Birds of prey, such as the buzzard, feed on thrushes.

Use this information to complete a food chain of the organisms involved.

Answer:

green plants → slugs → thrushes → birds of prey

Show you can ?

Explain two ways in which plants are essential for animal life on Earth.

Competition between living organisms

Living things compete with each other for resources.

For example, plants compete for:

▶ water
▶ light
▶ space to grow
▶ minerals.

Animals compete for:

▶ water
▶ food
▶ territory (space to live)
▶ mates.

Investigating competition

Competition between living organisms can be investigated by planting different numbers of seeds in flower pots and then measuring the mass of the seedlings (young plants) when they are grown. The graphs in Figure 8.7 show what usually happens.

In this type of experiment there are a number of things that need to be kept constant (controlled variables) for all the pots to make it a 'fair test' and give valid results.

These include:

▶ using the same size of pot
▶ using the same volume and type of compost
▶ ensuring that each pot gets the same amount of water
▶ placing the pots in the same conditions, for example, the same light and temperature.

Some species are such good competitors that they can harm other species. Some of these are discussed in the next section.

Competitive invasive species

Humans affect how living organisms compete in nature. One way they do this is by the introduction of new species into areas where they are not naturally found. This can be particularly harmful if competitive invasive species are introduced into new areas.

Competitive invasive species:

▶ out-compete similar native species, usually causing them harm
▶ spread rapidly when introduced into a region
▶ are almost always introduced to a country by humans (they are 'foreign').

Two examples are given below.

Grey squirrels

Grey squirrels are native to North America. They were introduced by humans to Ireland about 100 years ago. Now they have spread rapidly and out-compete the red squirrel because they are larger and can feed on a wider range of foods. They also are immune to a type of virus disease that can kill red squirrels. One way to prevent the red squirrel becoming extinct in Ireland is to keep some woodland areas free of grey squirrels.

a) when there are more plants in the pot none of the plants will grow as well – they are **competing** for resources such as water, nutrients and light

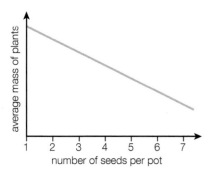

b) when there are more plants in the pot none of the plants will grow as well but because there are more plants the total mass will increase up to a limit when they become too cramped

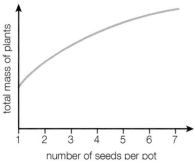

▲ **Figure 8.7** Competition in plants

△ **Figure 8.8** The red and the grey squirrel

Rhododendrons

The rhododendron is not a native British tree but was introduced as an ornamental plant. It spreads rapidly and, because it has leaves all year round, casts a dense shadow that prevents other types of plants living underneath it. Its effect can be reduced by removing young rhododendron shrubs as they grow.

Test yourself

7 State **two** resources that plants compete for.
8 State **two** resources that animals compete for.
9 Give **two** features that all competitive invasive species have.
10 Name **one** competitive invasive species.

△ **Figure 8.9** Rhododendron

Show you can (?)

Describe how you would set up an investigation into the effect of planting density on the average mass of seedlings in a pot.

Monitoring change in the environment

We can check how the environment is changing by observing both abiotic (non-living) and biotic (living) factors.

Abiotic (non-living) factors

We can monitor environmental change such as global warming by collecting information on:

▶ carbon dioxide levels
▶ size of ice fields and water levels
▶ temperature.

In addition, the pH of water can be used to monitor water pollution levels.

▲ **Figure 8.10** Lichens growing on trees in an area of unpolluted woodland

Tip

The less air pollution, the more lichen.

Biotic (living) factors

We can also monitor the health of the environment by studying the effect of, for example, pollution on living organisms. A good example is the use of lichens as pollution monitors.

Lichens are small, plant-like organisms that can be used to monitor pollution levels. The lichens grow on trees and rocks and grow best when there is no air pollution. A lot of air pollution prevents the growth of lichens.

How human activity can promote biodiversity

Biodiversity is the term used to describe the range of species (different types of living organisms) in a particular area. High biodiversity means that there are many species present.

For many years, human activity has harmed the world's biodiversity. This has occurred through clearing forests, pollution of air, sea and land, overfishing, global warming and for many other reasons.

However, in recent years there has been a greater understanding of the harmful effect that many human activities can have and attempts have been made to reduce this harm.

Agriculture

In the UK one of the major causes of decreasing biodiversity has been the effect of agriculture. Harmful effects include clearing woodland and hedgerows to provide more farmland and larger fields.

The overuse of fertiliser has allowed crop plants and grass for grazing to grow at the expense of other slower growing wild species.

Fertilisers running into waterways cause green algae to grow to excessive levels. When the algae die, bacteria decompose and break them down and use up most of the oxygen in the water, killing fish and other small animals.

Biodiversity in agricultural land can be increased by:

▶ replanting hedgerows
▶ keeping field margins 'wild' by not planting crops right to the edge of a field
▶ more efficient use of fertilisers so that there is not too much which can run off into waterways.

▲ **Figure 8.11** Virtually all the hedgerows have been removed from this agricultural land

Tip

The harmful effects of agriculture all lead to a decrease in biodiversity.

Tip

The crop plants outcompete the smaller and slower growing wild flowers as the fertiliser used has the function of causing the crop to grow as fast as possible – the fertiliser is suited to the crop.

Tip

The 'efficient' use of fertilisers can be achieved by measuring the mineral content of the soil before adding fertiliser – this will ensure only the right quantities are used and therefore there will not be an excess of fertiliser. Also, if farmers avoid spraying on wet days, less will be washed into surrounding rivers.

Better land use management

Better land use management can also help improve biodiversity.

The reclaiming and reuse of old industrial sites for industry avoids having to destroy existing habitats to provide more land. Also, the use of brownfield sites for housing prevents having to build on grassland or clearing woodland for housing, both of which would harm biodiversity by destroying habitats.

Additionally, planting new sustainable woodlands prevents having to cut down existing woodlands for timber.

Protecting fish stocks

Many of our favourite fish, such as cod and herring, are becoming rare because of overfishing. To stop these and other fish becoming extinct we must fish at a sustainable level. This means fishing at a level that does not cause the numbers to fall too much (Figure 8.12).

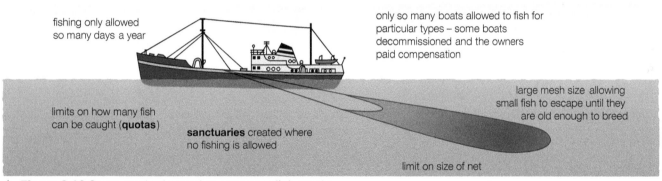

fishing only allowed so many days a year

only so many boats allowed to fish for particular types – some boats decommissioned and the owners paid compensation

large mesh size allowing small fish to escape until they are old enough to breed

limits on how many fish can be caught (**quotas**)

sanctuaries created where no fishing is allowed

limit on size of net

▲ **Figure 8.12** Some methods used to conserve fish stocks

Nature reserves

Nature reserves are also used to improve biodiversity in a region by protecting and conserving rare species and valuable or rare habitats.

Nature reserves can be open to the public or they can have no public access depending on how fragile the habitat is. As well as directly protecting rare plants and/or habitats, they can help promote biodiversity by educating the public about nature. Nature reserves are usually managed by government agencies or charities.

The role of international treaties in combating global pollution

Pollution that is on a worldwide scale such as global warming (climate change) requires a worldwide approach to combat it.

Global warming affects all countries so it is a problem that requires all countries to play their part. There have been two major treaties to attempt to reduce the rate of global warming.

The Kyoto Protocol in 1997 was a serious attempt at getting countries to do more to reduce fossil fuel emissions (the major factor in global warming). However, it had only limited success as it was based on voluntary agreements and many countries refused to support it.

The Paris Agreement (2015) has a target of limiting global warming to 2 °C compared to pre-industrial levels (and 1.5 °C if possible). It is also legally binding and many more countries have signed up.

Tip

A brownfield site is one that previously had housing (or other urban structures) on it.

Tip

A habitat is a place where organisms live. Habitats include woodland, grassland, a river, a garden, a field.

Test yourself

11 What is a biotic factor?
12 Define the term biodiversity.
13 What is a brownfield site?
14 Give two ways that we can preserve fish stocks.

Show you can

Explain one way in which the use of fertiliser can harm biodiversity.

Tip

A reduction in the use of fossil fuels is the most important step in reducing global warming.

1 a) i) Name the chemical that absorbs light energy in leaves. *(1 mark)*

   ii) Name the microscopic structures in plant leaves that contain this chemical. *(1 mark)*

b) The starch test is used to investigate if starch is present in leaves as a result of photosynthesis. The list below gives the steps for testing for starch but they are not in the correct order.

  A Add iodine to the leaf.

  B Remove a leaf from a plant and then place it in a beaker of boiling water for 30 seconds.

  C Dip the leaf into boiling water to soften it.

  D Spread the leaf flat on a white tile.

  E Boil the leaf in ethanol.

  i) Place the steps (A, B, C, D and E) in the correct order. *(2 marks)*

  ii) Give **one** safety precaution that should be taken when carrying out the starch test. *(1 mark)*

2 Pondweed gives off bubbles of oxygen gas when photosynthesising. It is possible to compare the rates of photosynthesis in different light intensities by comparing the number of bubbles released by the pondweed when placed in different light conditions.

Figure 8.13 shows how this type of investigation can be set up.

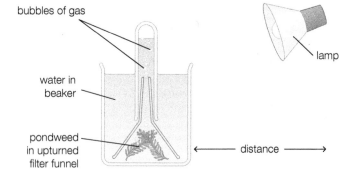

Figure 8.13

Typical results are shown in Table 8.2.

Table 8.2

Distance of lamp from the plant/cm	Number of bubbles/minute
5	46
10	28
15	10
20	9
25	9

a) Copy Figure 8.14, and then use the results to plot and draw a line graph on the grid. *(3 marks)*

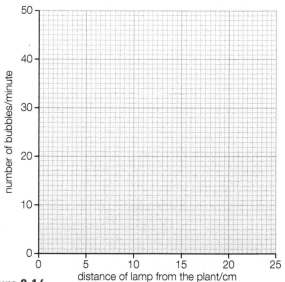

Figure 8.14

b) Describe and explain the results shown. *(3 marks)*

c) When carrying out this investigation at different light intensities, suggest **two** variables that should be kept the same. *(2 marks)*

d) i) Suggest **two** reasons why counting the number of bubbles given off over time may not be a very accurate way to compare rates of photosynthesis in different conditions. *(2 marks)*

  ii) Suggest a more accurate way to compare rates. *(1 mark)*

3 a) Figure 8.15 shows a simple food chain.

Figure 8.15

  i) Name the producer in this food chain. *(1 mark)*

  ii) Name the primary consumer in this food chain. *(1 mark)*

b) Figure 8.16 represents a grassland food web.

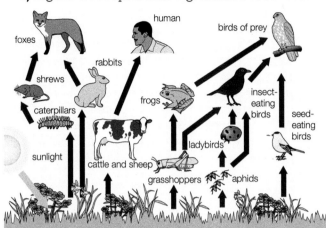

Figure 8.16

  i) Using the diagram of the food web, write out a food chain with five organisms. *(2 marks)*

  ii) Suggest why a food web is a more accurate way of showing feeding relationships than a food chain. *(1 mark)*

4 Figure 8.17 shows the relationship between the number of lichen plants on trees and their distance from a large city in 1970 and 2010.

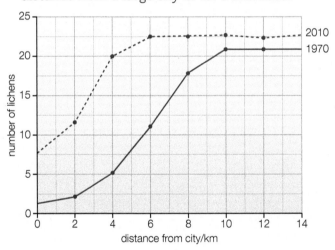

Figure 8.17

  a)i) Using the graph, describe and explain the results for the year 1970. *(3 marks)*

  ii) Suggest what has caused the change in lichen distribution between 1970 and 2010. *(1 mark)*

  b) When collecting the data, suggest **one** thing that must be done to make the investigation valid (keep it a 'fair test'). *(1 mark)*

5 a) Table 8.3 below shows the results from an investigation into how the number of plant seedlings in a pot affects the mass of the seedlings.

Table 8.3

Number of seedlings in each pot	Total mass of the seedlings/g	Average mass of each seedling/g
1	25	25
5	100	20
10	120	12
15	120	8
20	90	

  i) Calculate the missing value in the table. *(1 mark)*

  ii) Describe and explain how the number of seedlings in each pot affects the average mass of the seedlings. *(3 marks)*

  b) Competitive invasive species are species that have been introduced into a country by humans. Examples of competitive invasive species include the rhododendron and the grey squirrel. Give **two** other characteristics of all competitive invasive species. *(2 marks)*

6 a) Field margins are the areas around the edges of fields that are not used for crops, as shown in Figure 8.18.

crop   field margin   hedgerow

Figure 8.18

Scientists investigated how the width of field margins in fields affects the number of plant species found in the field margins. Their results are shown in Figure 8.19.

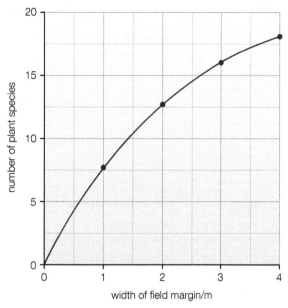

number of plant species

width of field margin/m

Figure 8.19

i) Describe the effect of field margin width on the number of plant species. *(1 mark)*

Scientists then investigated how the number of animal species in the field margins changed with increasing width. They found that the results were similar to those for the plants.

ii) Suggest **two** reasons for this trend in the number of animal species. *(2 marks)*

b) Nature reserves can also help increase biodiversity. Give **two** ways in which nature reserves can help promote biodiversity. *(2 marks)*

9 Acids, bases and salts

Specification points

This chapter covers sections 2.1.1 to 2.1.11 and 2.6.1 to 2.6.6. It is about acids, bases and salts, and includes the prescribed practical activities that demonstrate the reactions of these substances.

Acids, bases and salts

Safety in the laboratory

It is important to carry out any laboratory work in a safe manner. This would include identifying any risks associated with the experiment, taking steps to minimise these risks and being aware of what to do in case of an accident. With most chemistry experiments the main form of risk is from chemicals; therefore, these are labelled to identify the hazard associated with them.

Hazard labelling

The common hazard symbols shown in Table 9.1 are used to identify products that represent danger. Symbols have a greater visual impact than written warnings, and can be recognised easily and quickly by anyone, whatever language they speak. These same symbols are used around the world: they are internationally understood.

Many substances found in industry and around the home have hazard labels on them, for example:

▶ dynamite is explosive
▶ oven cleaner is corrosive
▶ weedkiller is toxic
▶ alcohol such as methylated spirit is flammable.

Tip

Make sure you can draw the hazard symbols, as this is a common exam question.

Table 9.1 Hazard symbols

Hazard symbols	Hazard
	Some acids and alkalis can burn your skin. Any substance that can burn skin is called corrosive.
	Some substances, including some acids, can poison you. Any substance that can poison you is called toxic.
	Some acids can be used to make substances that can cause an explosion. Any substance that can explode is called explosive.
	Some substances can catch fire easily. Any substance that can catch fire easily is called flammable.
	Some substances should be handled with care. For example, they may cause a skin irritation. Any substance that needs to be handled with care must be treated with caution.

▲ **Figure 9.1** Household substances that are acids or bases

Acids and bases

Acids are solutions that have a pH of less than 7 (see page 75). Soluble bases (alkalis) have a pH greater than 7. Acids react with bases in a reaction called neutralisation.

There are many acids and bases around your home, as shown in Figure 9.1. Table 9.2 shows some of the names of acids and alkalis found in everyday life.

Table 9.2 Some common acids and alkalis and their uses

Chemical name	Found in	Acid or alkali
Hydrochloric acid	Stomach acid	Strong acid
Sulfuric acid	Car batteries	Strong acid
Sodium hydroxide or potassium hydroxide	Oven cleaners and drain cleaners	Strong alkali
Ethanoic acid	Vinegar	Weak acid
Ammonia	Cleaning products	Weak alkali

A base that dissolves in water is also called an alkali. Sodium hydroxide is a base because it can neutralise an acid, but it also dissolves in water so it is an alkali as well as a base.

Tip

Remember: pH is always written with a small 'p' and a capital 'H'.

The pH scale

The strength of an acid or alkali is measured on the pH scale, which goes from 0 to 14.

Table 9.3 The pH scale

pH	0	1	2	3	4	5	6	7	8	9	10	11	12	13	14
Strength	Strong acid			Weak acid				Neutral	Weak alkali				Strong alkali		

Indicators

A colourless liquid may be neutral or it may be an acid or an alkali. A substance called an indicator can be used to distinguish between acids, alkalis and neutral solutions. It does this by changing colour.

Practical activity

Extracting an indicator from red cabbage

Some indicators occur naturally. Plants such as red cabbage, beetroot and blackcurrant contain dye (a coloured substance) that can act as an indicator.

⌃ **Figure 9.2** Making an indicator out of red cabbage

Procedure

1 Chop some red cabbage into small pieces or crush it using a pestle and mortar.

2 Place the pieces of red cabbage into a beaker and add some water.

3 Boil the mixture gently so that the colour comes out of the red cabbage into the water.

4 The water is now a solution of the indicator.

5 Allow the solution to cool, and then decant it (pour it off). You can do this through a filter funnel.

6 Discard the used red cabbage.

7 The liquid can now be used as an indicator.

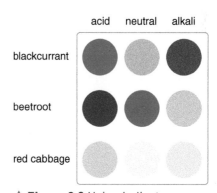

⌃ **Figure 9.3** Using indicators

This method can be used for any suitable plant. Figure 9.3 shows the colour changes that occur when a few drops of each indicator (red cabbage, beetroot or blackcurrant) are added to an acid, to water (neutral) and to an alkali.

Tip

If you are asked to give the method for making an indicator from a natural source, you must make it obvious that it is the coloured liquid that is used as the indicator and that the fruit/vegetable is discarded.

Litmus paper as an indicator

Litmus can be used as an indicator. It is one of the oldest forms of indicator, and is extracted from plant-like organisms called lichens. There are two different types: red litmus and blue litmus. Paper stained with litmus can be dipped into the test solution, and it then changes colour to indicate an acid or alkali (Figure 9.4). These colour changes are summarised in Table 9.4.

▲ **Figure 9.4** Colour changes of blue and red litmus

Table 9.4 Colour changes of red and blue litmus

	Acid	Neutral	Alkali
Red litmus	Red	Red	Blue
Blue litmus	Red	Blue	Blue

As you can see from Table 9.4, red litmus can only distinguish an alkali, and blue litmus can only distinguish an acid.

Universal indicator and pH

Although natural indicators can show whether a liquid is an acid or alkali, they cannot measure its strength. They do not show the difference between a strong acid, such as hydrochloric acid, and a weak acid, such as ethanoic acid in vinegar. You need to use an indicator with a larger range of colours such as universal indicator to show this.

Tip

Universal indicator is more useful than litmus paper because of the wider range of colours, which can tell us whether the acid or alkali is strong or weak.

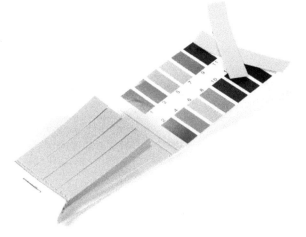

▲ **Figure 9.5** Colours of universal indicator

Universal indicator changes to a range of colours, depending on the strength of the acid or alkali. To use universal indicator, you need to put some of the solution you want to test into a test tube. Then add a few drops of universal indicator to it.

Universal indicator is a solution but you can also use **universal indicator paper**, which is paper that has been soaked in universal indicator and allowed to dry.

To test a solution, put a piece of universal indicator paper on a tile and then use a glass rod to put one drop of the solution onto the paper. The colour can be compared with a colour chart to work out the pH of the solution. Figure 9.6 shows sodium hydroxide being tested with universal indicator paper.

Table 9.5 summarises the colour changes of universal indicator in different substances and how these match to the pH scale.

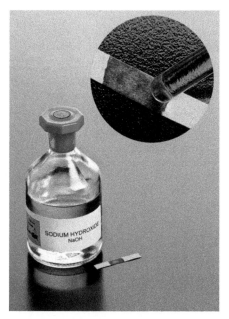

▲ **Figure 9.6** Testing sodium hydroxide with universal indicator paper

Table 9.5 Colours of universal indicator at different pH values

pH	0	1	2	3	4	5	6	7	8	9	10	11	12	13	14
Colour with universal indicator	Red			Orange		Yellow		Green	Green/blue		Blue		Purple		
Strength	Strong acid			Weak acid				Neutral	Weak alkali				Strong alkali		
Examples	Hydrochloric acid Sulfuric acid			Ethanoic acid				Water	Ammonia				Sodium hydroxide Potassium hydroxide		

Test yourself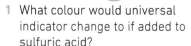

1 What colour would universal indicator change to if added to sulfuric acid?
2 Toothpaste is a weak alkali. What pH would best describe toothpaste?
3 What colour would red litmus change to if added to sodium hydroxide?
4 When universal indicator was added to a colourless liquid it turned green. What pH is the liquid?

Show you can

Table 9.6 shows the colour changes for two indicators.

Table 9.6 Colour changes for two indicators

Indicator	Colour at pH 0–6	Colour at pH 7	Colour at pH 8–14
A	Yellow	Green	Blue
B	Red	Orange	Orange

1 What colour will indicator A be in hydrochloric acid?
2 What colour will indicator B be in sodium hydroxide?
3 Which indicator (A or B) would be best for showing that a solution is neutral? Explain your answer.

▲ **Figure 9.7** A pH meter

Digitally measuring pH

You can also use a **pH sensor** or **pH probe** to measure pH. Attaching the sensor to a **data logger** allows the pH values to be transferred to a computer, and a graph of the change in pH value over time can be produced.

A **pH meter** can be used to measure pH accurately. It gives a numerical reading of pH to one or sometimes two decimal places. The one shown in Figure 9.7 is reading a pH of 3.12.

Neutralisation

When an acid reacts with an alkali, a neutralisation reaction occurs. If the alkali is added to the acid, the pH changes from a low value to a higher value as more alkali is added.

An acid is put in a beaker, with a pH sensor, and an alkali is added from a burette, as shown in Figure 9.8.

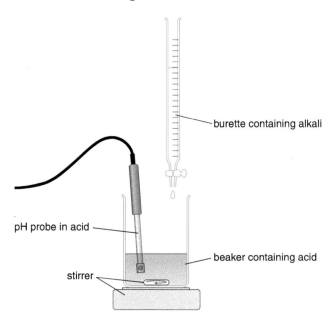

burette containing alkali

pH probe in acid

beaker containing acid

stirrer

▲ **Figure 9.8** A neutralisation reaction

A computer can draw a graph of pH against the volume of alkali added. If the acid is hydrochloric acid and the alkali is sodium hydroxide, the graph may look like that in Figure 9.9.

Tip

The neutralisation reaction graph is common in exams. You must be able to use the graph to work out the pH of the acid used and the pH of the alkali used in the reaction as well as the volume of acid or alkali added at the point of neutralisation (pH 7).

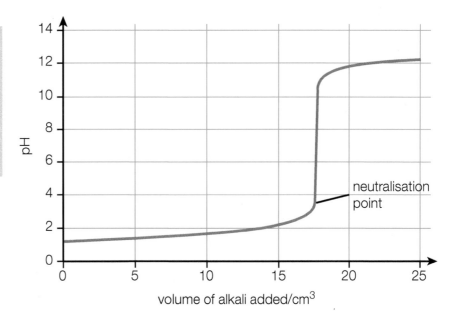

▲ **Figure 9.9** A neutralisation reaction graph

Tip

It is important to be familiar with the use of a burette. It is an important piece of apparatus in chemistry, and it is used because it is a more accurate way of measuring a volume of liquid than the simpler measuring cylinder.

A neutralisation reaction can also be studied using a chemical indicator such as universal indicator.

Add a few drops of universal indicator to the acid in the beaker (it will be red). Add the alkali slowly from the burette. The indicator will gradually change colour, and when the solution is neutral the indicator will be green.

Using data logging to record the data from a pH probe is much more accurate than trying to record the colour of an indicator as the neutralisation reaction takes place.

Prescribed practical

Prescribed practical C1: Following a neutralisation reaction by monitoring the pH

Procedure

1 You will be given an acid and an alkali.

2 Fill a burette with the alkali.

3 Measure out 20 cm³ of the acid and place it in a conical flask or beaker.

4 Add a few drops of universal indictor to the alkali or insert a pH sensor/meter.

5 Record the pH.

6 Start adding 2 cm³ of the alkali at a time and record the pH of the solution.

7 Continue adding the alkali until you reach a pH of 9 or more.

8 Copy and complete Table 9.7 below (add extra columns if required).

Table 9.7 Recording pH as an alkali is added to an acid

Volume of alkali added/ cm³	0	2	4	6	8	10	12	14	16	18	20	22	24	26
pH														

9 Draw a graph of pH against the volume of alkali added, using axes similar to the outline drawn in Figure 9.10. When drawing this graph you should use a smooth curve (not a point-to-point graph).

10 Describe and explain your results.

You can also complete a similar investigation by adding an acid to an alkali. The initial pH would start at a high pH value and end at a lower one.

Questions and sample data

1 Explain how you would use your graph to find the volume of alkali added to neutralise the acid.

2 Explain why using a pH probe or pH meter is better than using universal indicator to follow the reaction.

3 A student carried out a similar reaction. He added an acid to an alkali. His results are shown in Table 9.8.

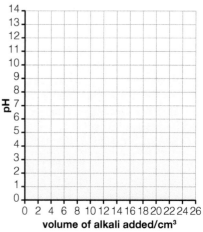

▲ **Figure 9.10** Axes to plot a neutralisation curve

Table 9.8 Results of a neutralisation reaction

Volume of acid added/cm³	0	2	4	6	8	10	12	14	16	18
pH	12	12	12	11.5	11	7	3	2.5	2.5	2.5

a) How much acid had he added when the solution was neutral?

b) How can you tell he used a pH probe to record the pH rather than an indicator?

c) Why did he stop adding acid at 18 cm³?

d) What was the pH of the acid and alkali that he used?

e) If he repeated the reaction with a stronger acid, what prediction could you make about the point of neutralisation?

▲ **Figure 9.11** The indigestion remedy '*milk of magnesia*' contains magnesium hydroxide

Everyday examples of neutralisation

Acid indigestion

A neutralisation reaction can be useful. For example, your stomach contains hydrochloric acid, which helps prevent infection and digest food. Sometimes, though, there can be too much acid in your stomach. This **excess** acid causes acid indigestion. An indigestion remedy such as baking soda (sodium hydrogencarbonate) can neutralise some of the stomach acid. Substances that are used to neutralise excess hydrochloric acid in the stomach are called antacids. The indigestion remedy called '*milk of magnesia*' (see Figure 9.11) contains **magnesium hydroxide**. The main antacids are magnesium hydroxide, calcium carbonate and sodium hydrogencarbonate.

Toothpaste

Many of the foods we eat contain weak acids. Also, bacteria in the mouth produce acid as they break down any sugars from the foods we eat. When the pH of the mouth is 5.5 or less, tooth decay occurs. By brushing teeth with a toothpaste that is an alkali, this acid can be neutralised and so reduce tooth decay.

Baking soda (sodium hydrogencarbonate), which is sometimes called bicarbonate of soda, is often added to toothpaste to neutralise acid in the mouth.

General reactions of acids

The most common acids used in a chemistry laboratory are sulfuric acid and hydrochloric acid.

When an acid reacts with a metal or a metal compound, a salt is formed along with some other products, such as water or hydrogen. We can summarise a chemical reaction by using a word equation.

Table 9.9 summarises the reactions of acids that you may be asked about:

Table 9.9 The reactions of acids

Reactants	Products	Observations
acid + metal	salt + hydrogen	Bubbles of gas given off, gets warmer
acid + base/alkali	salt + water	Gets warmer
acid + metal carbonate	salt + water + carbon dioxide	Bubbles of gas given off, gets warmer
acid + metal hydrogencarbonate	salt + water + carbon dioxide	Bubbles of gas given off, gets warmer

Example

1 Hydrochloric acid in the stomach can react with the magnesium hydroxide in an antacid. During this reaction, magnesium chloride (a salt) is formed along with water. The word equation is:

hydrochloric acid + magnesium hydroxide → magnesium chloride + water

2 If sulfuric acid reacted with magnesium hydroxide, a different salt (magnesium sulfate) would be formed along with water. The word equation is:

sulfuric acid + magnesium hydroxide → magnesium sulfate + water

Naming salts

The name of the salt produced depends on the acid and metal or metal compound being used.

The first part of the name of the salt comes from the metal involved, while the second part comes from the acid being used.

▶ When hydrochloric acid is used, a chloride is formed.

▶ When sulfuric acid is used, a sulfate is formed.

> **Example**
>
> 1 hydrochloric acid + sodium hydroxide → sodium chloride + water
>
> 2 sulfuric acid + magnesium hydroxide → magnesium sulfate + water
>
> 3 ethanoic acid + calcium carbonate → calcium ethanoate + water + carbon dioxide
>
> In the third example, **carbon dioxide** is also formed because a metal **carbonate** is used.

Testing for gases

If a gas is produced during a reaction and collected, it can be tested to identify it.

Test for hydrogen

Hydrogen is explosive, but making it in small quantities is safe. A lighted splint will burn with a 'squeaky pop' in the presence of hydrogen, as shown in Figure 9.12.

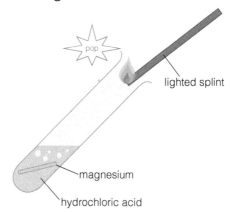

pop

lighted splint

magnesium

hydrochloric acid

▲ **Figure 9.12** Testing for hydrogen gas

The equation for this reaction is:

hydrogen + oxygen → water

$$2H_2 + O_2 \rightarrow 2H_2O$$

Test for carbon dioxide

Limewater (calcium hydroxide solution) is used to test for the presence of carbon dioxide, as shown in Figure 9.13. It is a colourless liquid that turns cloudy or milky in the presence of carbon dioxide. Other gases have no effect on it.

> **Tip**
>
> Many students write the word 'clear' instead of colourless. This is not acceptable as an exam answer.

Tip

If asked for the observations when using limewater to test for carbon dioxide, it is important to note both the starting colour (colourless) and the end colour (milky or cloudy).

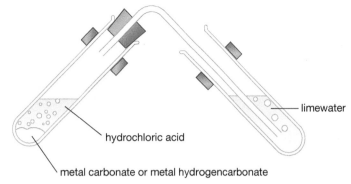

▲ **Figure 9.13** Testing for carbon dioxide gas

Test for oxygen

Although not produced in any of the reactions with acid described, oxygen is another colourless gas. It can be tested for using a glowing splint, which will relight in the presence of oxygen, shown in Figure 9.14.

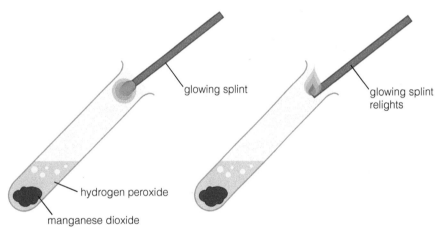

▲ **Figure 9.14** Testing for oxygen gas

Tip

An element's symbol will be only one or two letters (some elements have three letter symbols, though these are not covered in the specification). If it is a single letter, then it must be a capital letter, for example carbon is C and potassium is K. If there are two letters, then the first will be a capital and the second a lower-case letter: for example, calcium is Ca and iron is Fe. Incorrect symbols will not gain marks in an exam question.

Chemical formulae and balancing equations

You will encounter many chemical formulae in this chemistry unit. These can represent elements and compounds (see Chapter 10). It is important that you are familiar with them and can recognise and write them accurately. The symbols for the elements are given in the Periodic Table.

A single symbol represents an atom of the element. Na represents an atom of sodium, O represents an atom of oxygen, H represents an atom of hydrogen and Cl represents an atom of chlorine.

Examples of writing word and symbol equations follow.

1 The reaction of magnesium with hydrochloric acid

hydrochloric acid + magnesium → magnesium chloride + hydrogen

In this example the chemical formulae are as follows:

- hydrochloric acid is HCl

- magnesium is Mg

- magnesium chloride is $MgCl_2$

- hydrogen is H_2.

For any formula, it is important that the correct symbols for the elements are used, with correct capital and lower-case letters. The numbers are subscript.

- Hydrochloric acid (HCl) contains atoms of two different elements – one atom of hydrogen (H) and one atom of chlorine (Cl).

- Magnesium chloride ($MgCl_2$) contains atoms of two different elements – one atom of magnesium (Mg) needs two atoms of chlorine (Cl) to form a compound.

- Hydrogen (H_2) contains two atoms that are the same – both hydrogen (H). The small '2' after the symbol means two atoms.

The **symbol equation** is therefore:

$HCl + Mg → MgCl_2 + H_2$

It is also important that this symbol equation is balanced, which means that there are the same number of atoms of each element on each side of the equation. (This is because elements or atoms cannot be made or destroyed during a reaction – they can only be rearranged.)

HCl + Mg		→	MgCl$_2$ + H$_2$	
Left-hand side:	H: 1 Cl: 1 Mg: 1		Right-hand side:	H: 2 Cl: 2 Mg: 1

In this example, the equation is not balanced: there are double the number of hydrogen and chlorine atoms on the right-hand side compared with the left-hand side. This can be corrected by having two molecules of hydrochloric acid on the left-hand side:

2HCl + Mg		→	MgCl$_2$ + H$_2$	
Left-hand side:	H: ~~1~~ **2** Cl: ~~1~~ **2** Mg: 1		Right-hand side:	H: 2 Cl: 2 Mg: 1

2 The reaction of magnesium hydroxide with sulfuric acid

   magnesium hydroxide + sulfuric acid $\rightarrow$ magnesium sulfate + water

In this example, the chemical formulae are as follows:

- magnesium hydroxide is $Mg(OH)_2$

- sulfuric acid is H_2SO_4

- magnesium sulfate is $MgSO_4$

- water is H_2O.

For any formula, it is important that the correct symbols for the elements are used, with correct capital and lower-case letters. The numbers are subscripts.

- The formula for magnesium hydroxide ($Mg(OH)_2$) contains atoms of three different elements – one atom of magnesium (Mg), two atoms of oxygen (O) and two atoms of hydrogen (H). (The brackets around 'OH' mean that there are two of each element inside the brackets).

- Sulfuric acid (H_2SO_4) contains atoms of three different elements – two atoms of hydrogen (H), one atom of sulfur (S) and four atoms of oxygen (O).

- Magnesium sulfate ($MgSO_4$) contains atoms of three different elements – one atom of magnesium (Mg), one atom of sulfur (S) and four atoms of oxygen (O).

- Water contains atoms of two different elements – two atoms of hydrogen (H) and one atom of oxygen (O).

The **symbol equation** is therefore:

$$Mg(OH)_2 + H_2SO_4 \rightarrow MgSO_4 + H_2O$$

It is also important that this symbol equation is balanced, which means that there are the same number of atoms of each element on each side of the equation.

$Mg(OH)_2 + H_2SO_4$		$\rightarrow$	$MgSO_4 + H_2O$	
Left-hand side:	Mg: 1		Right-hand side:	Mg: 1
	O: 6			O: 5
	H: 4			H: 2
	S: 1			S: 1

In this example, the equation is not balanced: there are double the number of hydrogen atoms on the left-hand side compared with the right-hand side and one more oxygen atom on the left-hand side. This can be corrected by having two molecules of water on the right-hand side:

$Mg(OH)_2 + H_2SO_4$		$\rightarrow$	$MgSO_4 + 2H_2O$	
Left-hand side:	Mg: 1		Right-hand side:	Mg: 1
	O: 6			O: ~~5~~ 6
	H: 4			H: ~~2~~ 4
	S: 1			S: 1

3 The reaction of hydrochloric acid with calcium carbonate

hydrochloric acid + calcium carbonate → calcium chloride + water + carbon dioxide

In this example, the chemical formulae are as follows:

- hydrochloric acid is HCl
- calcium carbonate is $CaCO_3$
- calcium chloride is $CaCl_2$
- water is H_2O
- carbon dioxide is CO_2.

For any formula, it is important that the correct symbols for the elements are used, with correct capital and lower-case letters. The numbers are subscripts.

- The formula for hydrochloric acid (HCl) contains atoms of two different elements – one atom of hydrogen (H) and one atom of chlorine (Cl).
- Calcium carbonate ($CaCO_3$) contains atoms of three different elements – one atom of calcium (Ca), one atom of carbon (C) and three atoms of oxygen (O).
- Calcium chloride ($CaCl_2$) contains atoms of two different elements – one atom of calcium (Ca) and two atoms of chlorine (Cl).
- Water contains atoms of two different elements – two atoms of hydrogen (H) and one atom of oxygen (O).
- Carbon dioxide (CO_2) contains atoms of two different elements – one atom of carbon (C) and two atoms of oxygen (O).

The **symbol equation** is therefore:

$$HCl + CaCO_3 \rightarrow CaCl_2 + H_2O + CO_2$$

It is also important that this symbol equation is balanced, which means that there are the same number of each element on each side of the equation.

$HCl + CaCO_3$		→	$CaCl_2 + H_2O + CO_2$	
Left-hand side:	H: 1		Right-hand side:	H: 2
	Cl: 1			Cl: 2
	Ca: 1			Ca: 1
	C: 1			C: 1
	O: 3			O: 3

In this example, the equation is not balanced: there are double the number of hydrogen and chlorine atoms on the right-hand side compared with the left-hand side. This can be corrected by having two molecules of hydrochloric acid on the left-hand side:

$2HCl + CaCO_3$		$\rightarrow$	$CaCl_2 + H_2O + CO_2$	
Left-hand side:	H: ~~1~~ **2**		Right-hand side:	H: 2
	Cl: ~~1~~ **2**			Cl: 2
	Ca: 1			Ca: 1
	C: 1			C: 1
	O: 3			O: 3

4 The reaction of sulfuric acid with sodium hydrogencarbonate

sulfuric acid + sodium hydrogencarbonate $\rightarrow$ sodium sulfate + water + carbon dioxide

In this example, the chemical formulae are as follows:

- sulfuric acid is H_2SO_4
- sodium hydrogencarbonate is $NaHCO_3$
- sodium sulfate is Na_2SO_4
- water is H_2O
- carbon dioxide is CO_2.

For any formula, it is important that the correct symbols for the elements are used, with correct capital and lower-case letters. The numbers are subscripts.

- The formula for sulfuric acid (H_2SO_4) contains atoms of three different elements – two atoms of hydrogen (H), one atom of sulfur (S) and four atoms of oxygen (O).
- Sodium hydrogencarbonate ($NaHCO_3$) contains atoms of four different elements – one atom of sodium (Na), one atom of hydrogen (H), one atom of carbon (C) and three atoms of oxygen (O).
- Sodium sulfate (Na_2SO_4) contains atoms of three different elements – two atoms of sodium (Na), one atom of sulfur (S) and four atoms of oxygen (O).
- Water contains atoms of two different elements – two atoms of hydrogen (H) and one atom of oxygen (O).
- Carbon dioxide (CO_2) contains two different elements – one atom of carbon (C) and two atoms of oxygen (O).

The **symbol equation** is therefore:

$$H_2SO_4 + NaHCO_3 \rightarrow Na_2SO_4 + H_2O + CO_2$$

It is also important that this symbol equation is balanced, which means that there are the same number of each element on each side of the equation.

$H_2SO_4 + NaHCO_3$		$\rightarrow$	$Na_2SO_4 + H_2O + CO_2$	
Left-hand side:	H: 3		Right-hand side:	H: 2
	S: 1			S: 1
	O: 7			O: 7
	Na: 1			Na: 2
	C: 1			C: 1

In this example, the equation is not balanced: there are double the number of sodium atoms on the right-hand side compared with the left-hand side, and there is also one more hydrogen atom on the left-hand side. This can be corrected by having two molecules of water and carbon dioxide on the right-hand side and two of sodium hydrogencarbonate on the left-hand side.

$H_2SO_4 + \mathbf{2}NaHCO_3$		$\rightarrow$	$Na_2SO_4 + \mathbf{2}H_2O + \mathbf{2}CO_2$	
Left-hand side:	H: ~~3~~ **4**		Right-hand side:	H: ~~2~~ **4**
	S: 1			S: 1
	O: ~~7~~ **10**			O: ~~7~~ **10**
	Na: ~~1~~ **2**			Na: 2
	C: ~~1~~ **2**			C: ~~1~~ **2**

Test yourself

5 a) Explain what is meant by the term 'neutralisation reaction'.
  b) Give an example of neutralisation in everyday life.
6 Copy and complete the word equations for the following acid reactions:
  a) ethanoic acid + calcium →
  b) sulfuric acid + sodium hydroxide →
  c) hydrochloric acid + magnesium carbonate →
7 Copy and complete Table 9.10 for the different gas tests.

Table 9.10 Gas tests

Gas	Method	Result
Hydrogen		It burns with a squeaky pop
	Bubble gas through limewater	
Oxygen	Use a glowing splint	

8 Describe the observations for the reaction between sodium carbonate and sulfuric acid.
9 Copy and complete the balanced symbol equations for the following acid reactions:
  a) HCl + Ca →
  b) H_2SO_4 + NaOH →
  c) HCl + $MgCO_3$ →

Show you can ?

The salt calcium chloride can be made by reacting hydrochloric acid with calcium, calcium carbonate or calcium hydroxide. Copy and complete the diagram to show the products for each of these reactions.

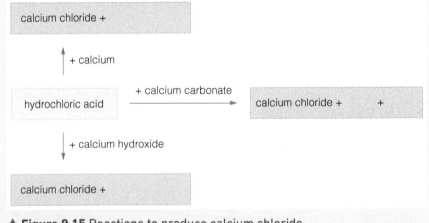

▲ **Figure 9.15** Reactions to produce calcium chloride

Practice questions

1 Table 9.11 gives information about some household acids and alkalis.
   a) Copy and complete the table. *(3 marks)*

Table 9.11

Substance	pH	Colour with universal indicator	Type of solution
Toothpaste		Green/blue	Weak alkali
Oven cleaner	13	Purple	
Vinegar	3		Strong acid

   b) How is a solution with pH 7 described? *(1 mark)*
   c) i) Draw the hazard symbol you would expect to see on a strong acid such as hydrochloric acid. *(1 mark)*
      ii) Name the hazard symbol found on a bottle of strong acid. *(1 mark)*
   d) Give **two** reasons why hazard symbols and not just words are put on bottles of chemicals. *(2 marks)*

2 a) What is meant by the term 'indicator'? *(1 mark)*
   b) Indicators can be made from various fruit and vegetables. Describe a method that could be used to extract an indicator from beetroot. *(3 marks)*
   c) Table 9.12 shows the colour changes of beetroot at various pH values.

Table 9.12

pH	2	5	7	10	13
Colour of beetroot indicator	Dark purple	Dark purple	Red	Light green	Light green

   i) What colour would beetroot indicator be in hydrochloric acid? *(1 mark)*
   ii) What colour would beetroot indicator be in ammonia? *(1 mark)*
   iii) Why would beetroot indicator not be suitable to distinguish between hydrochloric acid and ethanoic acid (vinegar)? *(1 mark)*
   d) What is the most accurate method to measure the pH of a solution? *(1 mark)*

3 A student wanted to follow the pH changes during the reaction between hydrochloric acid and sodium hydroxide. He used a pH meter. The graph in Figure 9.15 was obtained.

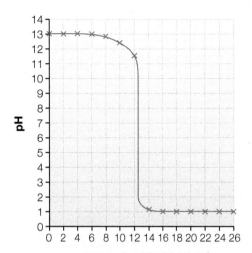

Figure 9.16
   a) What was the pH value of the sodium hydroxide at the start of this experiment? *(1 mark)*
   b) What volume of acid was needed to cause a sudden drop in the pH value? *(1 mark)*
   c) Name a suitable piece of apparatus that could have been used to add the acid during this experiment. *(1 mark)*
   d) i) Write a word equation for the reaction between hydrochloric acid and sodium hydroxide. *(2 marks)*
      ii) Write a balanced symbol equation for the reaction between hydrochloric acid and sodium hydroxide. *(2 marks)*
   e) Explain fully why a pH meter is better than a chemical indicator to follow this reaction. *(2 marks)*

4 Carbon dioxide can be made when hydrochloric acid is added to calcium carbonate.
   a) What name is given to this type of reaction? *(1 mark)*
   b) Copy and complete the word equation for this reaction: *(2 marks)*

   hydrochloric acid + calcium carbonate → carbon dioxide + _____ + _____

   c) i) Name the chemical used to test for carbon dioxide. *(1 mark)*
      ii) Describe the colour change during the test for carbon dioxide. *(2 marks)*

5 A student carried out a neutralisation reaction using hydrochloric acid and potassium hydroxide. She recorded the pH as she added the acid to the alkali. Her results are shown in Table 9.13.

Table 9.13

Volume of alkali added/ cm³	0	2	4	6	8	10	12	14	16
pH		12.0	12.0	11.5	11.3	3.0	2.7	2.5	2.5	2.5

a) These results were used to draw a graph of pH against the volume of alkali. Sketch the shape of the graph you would expect. *(2 marks)*

b) What was the pH of:
   i) the alkali?
   ii) the acid? *(2 marks)*

c) Suggest a piece of apparatus that could have been used to measure the volume of alkali. Give a reason for your choice. *(2 marks)*

d) Was the reaction followed using universal indicator paper or a pH probe? Give a reason for your choice. *(2 marks)*

e) Suggest why the student would have swirled the flask as she added the alkali to the acid. *(1 mark)*

f) i) Write a word equation for the reaction between hydrochloric acid and potassium hydroxide. *(2 marks)*
   ii) Write a balanced symbol equation for the reaction between hydrochloric acid and potassium hydroxide. *(2 marks)*

g) If the student wanted to make sure that her results were reliable, state one thing she could do. *(1 mark)*

h) The student also measured the temperature during the reaction. What would you expect to happen to the temperature during the neutralisation reaction? *(2 marks)*

10 Elements, compounds and mixtures

Solids, liquids and gases

All substances can exist in three different states: solids, liquids and gases. For example, water is a **liquid**: if it is cold enough (less than 0 °C) it will turn into ice, a **solid**; if it is hot enough (more than 100 °C) it will turn into steam, a **gas**. The three states of water are shown in FIgure 10.1.

▲ **Figure 10.1** The three states of water

All substances contain **particles**. The way in which the particles are arranged determines whether a substance is a solid, a liquid or a gas.

Figure 10.2 is a simple model to show how the particles are arranged in solids, liquids and gases.

Tip

Make sure you can accurately draw the particle arrangements, as this is a common exam question.

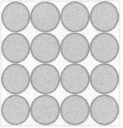
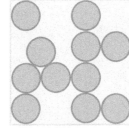
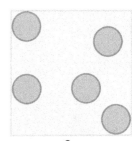

Solid	Liquid	Gas
Particles are very close together in a regular arrangement and are held in place by strong forces.	Particles are very close together, with most touching each other, and are in a more random arrangement.	Particles move around rapidly all the time and are very spread out.
Particles cannot move, but they do vibrate.	Particles can move over each other.	Gases have no fixed shape or volume. They take the shape and volume of the container they are in.
Solids have a fixed shape and volume.	Liquids can flow and take the shape of the bottom of the container. They have a fixed volume.	Gases can be easily compressed (the particles pushed closer together).

⋏ **Figure 10.2** Particle diagrams for solids, liquids and gases

Changes of state

A substance changes state because energy (heat) is either added to it or taken away from it.

The processes involved in changing state are shown in Figure 10.3.

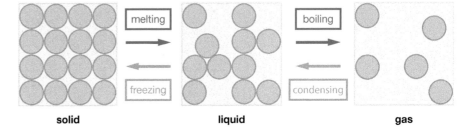

⋏ **Figure 10.3** Changes of state

To change from solid to liquid to gas, energy needs to be added (the substance is heated).

To change from gas to liquid to solid, energy needs to be removed (the substance is cooled).

The term **evaporating** can be used to replace boiling. However, when a substance evaporates, only the particles on the surface of the liquid are involved, and the liquid will not bubble. Evaporation can happen at all temperatures. When a substance boils, there is enough energy for all the particles to be involved, and the liquid will bubble. Boiling only happens at a fixed temperature (the boiling point).

Some substances can be changed from a solid straight into a gas (with no liquid state) or vice versa. This process is called **subliming**.

Iodine sublimes easily from a grey/black solid into a purple gas, as shown in Figure 10.4.

Solids, liquids and gases

▲ **Figure 10.4** Iodine subliming

Tip

It is important to remember that the term 'subliming' is used to describe two changes of state: from solid to gas and from gas back to solid.

Changing state is a physical reaction. This means that it can be reversed: for example, water that has become ice can easily be changed back into water by heating it.

State symbols

State symbols (shown in Table 10.1) can be added to balanced symbol equations to show whether each substance is a solid, liquid or gas. A fourth state symbol indicates whether the substance is dissolved in water, and is described as being aqueous. The state symbol is put after the substance, in brackets.

Table 10.1 State symbols

State	State symbol
solid	(s)
liquid	(l)
gas	(g)
aqueous (dissolved in water)	(aq)

For example, the equation for sodium hydroxide reacting with hydrochloric acid would be:

$$NaOH(aq) + HCl(aq) \rightarrow NaCl(aq) + H_2O(l)$$

Melting points and boiling points

Every substance has a **unique** melting point and boiling point, as shown in Table 10.2.

The melting point is the temperature at which a solid turns into a liquid.

The boiling point is the temperature at which a liquid turns into a gas.

Table 10.2 Melting and boiling points of different substances

Substance	Melting point/°C	Boiling point/°C
water	0	100
mercury	−39	360
sulfur	113	445
chlorine	−101	−35
copper	1083	2567
oxygen	−218	−183

A substance will be a solid below its melting point. Between its melting and boiling point it will be a liquid. Above the boiling point it will be a gas.

Consider water, for example. The melting point of water is 0 °C and its boiling point is 100 °C. Below 0 °C water will be a solid (ice), between 0 °C and 100 °C it will be a liquid and above 100 °C water will be a gas (steam), as shown in Figure 10.5.

93

H Show you can

What state are each of the substances shown in Table 10.3 at room temperature (25 °C)? Use Table 10.2 to help.

Table 10.3

Substance	State
Water	
Mercury	
Sulfur	
Chlorine	
Copper	
Oxygen	

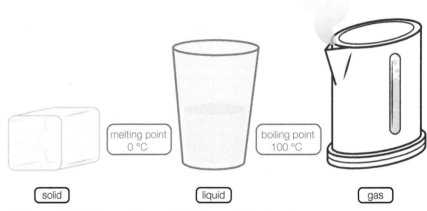

▲ **Figure 10.5** The melting and boiling points of water

At room temperature (approximately 25 °C), water will be a liquid because this temperature (25 °C) falls in between its melting and boiling points.

Test yourself

1 a) Name the three states of matter.
   b) Define the terms:
      i) freezing
      ii) condensing
      iii) subliming.
   c) Draw a particle diagram to represent a solid.
   d) Describe what happens to the arrangement of particles as a substance turns from a solid into a liquid.
2 Melting ice is a physical reaction.
   a) What is meant by the term 'physical reaction'?
   b) Give another example of a physical reaction.
3 a) Define the term 'boiling point'.
   b) What is the boiling point of water?

Elements, compounds and mixtures

All substances are elements, compounds or mixtures. They are made up of particles called atoms.

The atoms of any one element are all identical. For example, all carbon atoms are the same type of atom and all magnesium atoms are the same type of atom.

The atoms of different elements are different. For example, an atom of sodium is different from an atom of helium.

The way in which these atoms are arranged or joined in a substance determines whether it is described as an element, a compound or a mixture.

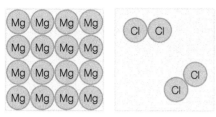

▲ **Figure 10.6** Atoms in magnesium and chlorine

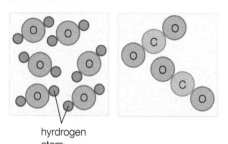

hyrdrogen atom

▲ **Figure 10.7** Atoms in water and carbon dioxide

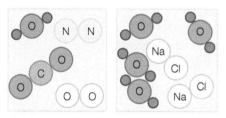

▲ **Figure 10.8** Atoms in air and sea water

Elements

Elements are only made up of one type of atom – they cannot be broken down into anything simpler.

For example, magnesium (Mg) is an element, as it only contains one type of atom – magnesium atoms. Chlorine (Cl_2) is also an element, as it only contains one type of atom – chlorine atoms (see Figure 10.6). All elements can be found in the Periodic Table (see Chapter 11).

Compounds

Compounds are made up of two or more different types of atom chemically joined together – they cannot be separated easily.

For example, water (H_2O) is a compound, as it contains two types of atom chemically joined – two hydrogen (H) atoms and an oxygen (O) atom. Carbon dioxide (CO_2) is also a compound, as it contains two types of atom chemically joined – one carbon (C) atom and two oxygen (O) atoms. Figure 10.7 shows atoms in water and carbon dioxide.

Mixtures

A mixture is when there are two or more substances – these may be elements or compounds – that are together but not chemically joined. Mixtures are relatively easy to separate.

For example, air is a mixture of different gases (including the elements nitrogen (N_2) and oxygen (O_2) and the compounds carbon dioxide (CO_2) and water vapour (H_2O). Sea water is a mixture of two compounds – water (H_2O) and sodium chloride (NaCl). Figure 10.8 shows atoms in air and sea water.

Test yourself

4 Define the terms:
   a) element
   b) compound
   c) mixture.
5 Copy and complete Table 10.4 by adding a cross to identify elements, compounds and mixtures. The first one has been done for you.

Table 10.4 Elements, compounds and mixtures

Substance	Element	Compound	Mixture
Silver	✗		
Water			
Oxygen			
Salt (sodium chloride)			
Sea water			
Carbon dioxide			
Nitrogen			
Air			

Show you can ❓

For each of the particle diagrams a), b), c) and d) in Figure 10.9, decide whether they are elements, compounds or mixtures. Give a reason for your answers.

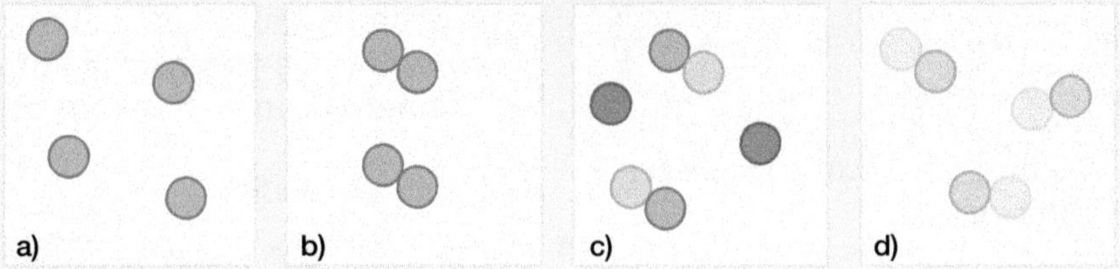

▲ **Figure 10.9** Types of substance

Pure and impure substances

Pure substances contain a single element or compound that is not mixed with any other substance. A pure substance will have a definite melting point and boiling point.

If there are only small amounts of other substances, these are called **impurities.** Impure substances will melt and boil over a range of temperatures. For example, petrol is a mixture of substances, and it boils over a range of temperatures from about 60 °C to 100 °C.

Separating mixtures

By definition, mixtures are relatively easy to separate in the laboratory. Separation methods include:

▶ filtration

▶ evaporation or crystallisation

▶ simple distillation

▶ paper chromatography.

Separating mixtures of solids and liquids

When a solid is mixed with a liquid, two things can happen: the solid will either dissolve or not dissolve.

If the solid dissolves, for example salt, it is described as being soluble. If the solid does not dissolve, for example sand, it is described as being insoluble.

When a solid dissolves in a liquid, the solid is called the solute and the liquid is called the solvent. Together, the solute and solvent make a solution.

For example, when salt (solute) dissolves in water (solvent), a salt solution forms.

Example ←

Table 10.5

Solute (soluble solid)	+	Solvent (liquid)	→	Solution
Salt	+	Water	→	Salty water

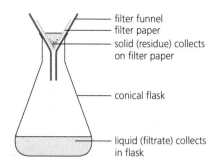

▲ **Figure 10.10** Filtration

▲ **Figure 10.11** Evaporation

Tip

Make sure that you practise drawing the apparatus set-up for filtration and evaporation. Use a pencil and ruler and add labels for the key pieces of apparatus.

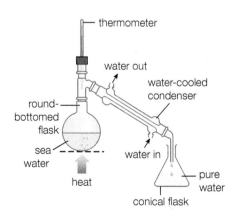

▲ **Figure 10.13** Distillation of sea water

Filtration

An insoluble solid and a liquid, for example a mixture of sand and water, can be separated using filtration. The apparatus for filtration is a **filter funnel** and **filter paper**, as shown in Figure 10.10.

The filter paper is folded and placed inside the filter funnel, and the mixture is then poured through the paper and funnel. The liquid will pass through the paper and funnel: it is called the filtrate. The insoluble part remains in the filter paper: it is called the residue.

Evaporation and crystallisation

A soluble solid and a liquid, for example a solution of salt and water, can be separated using **evaporation**. The apparatus for evaporation is an **evaporating dish** and a **Bunsen burner** (with a tripod and gauze), as shown in Figure 10.11. The mixture is placed in the evaporating dish and is heated until the liquid (solvent) has evaporated or boiled. The solid is left behind in the evaporating dish, and is called the **residue**.

Crystallisation is also used to separate a dissolved solid from a solvent. For example, it can be used to separate copper sulfate crystals from a copper sulfate solution. The mixture is heated to boil off some of the solvent, and, as it cools down, some of the solute crystallises out of the solution as crystals, as shown in Figure 10.12. If required, the rest of the solvent can be removed using filtration.

▲ **Figure 10.12** Crystallisation of copper sulfate

Simple distillation

A solvent can be separated from a solution, for example, water can be collected from sea water, using simple distillation. The apparatus for simple distillation is a (Liebig) condenser, a round-bottomed flask and a Bunsen burner (see Figure 10.13). The mixture is heated until the solvent boils, and the vapour produced passes through the condenser, where it cools and condenses. The condensed solvent is then collected, and is called the distillate.

Simple distillation can also be used to separate two miscible liquids such as alcohol and water. Miscible liquids are liquids that are able to mix together. Simple distillation works because the liquids have two different **boiling points**. The liquid with the lowest boiling point boils first and is collected as the distillate.

H

Paper chromatography

There are lots of different types of chromatography. **Paper chromatography** is used to separate mixtures of substances dissolved in a solvent. For example, the different coloured dyes in ink can be separated using paper chromatography.

Practical activity

Carrying out paper chromatography

1 Draw a pencil line about 1 cm from the bottom of the chromatography paper. (Pencil is used as it will not dissolve in the solvent or interfere with the results.)

2 Place a small spot of each of the test substances onto the pencil line and label them.

3 Place or hang the paper in a beaker with a small amount of solvent at the bottom (less than 1 cm so it will not touch the spots on the pencil line – this prevents the spots dissolving into the solvent).

4 Leave for a few minutes so that the solvent will travel up the chromatography paper.

5 When the solvent is near the top of the paper, take the paper out of the solvent and mark the level that the solvent reached: this is called the solvent front.

6 Leave to dry. This paper with marks from the substances is now called a chromatogram.

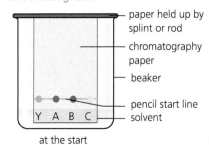
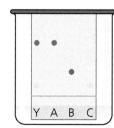

at the start   after the solvent has soaked up the paper

▲ **Figure 10.14** Paper chromatography

In paper chromatography, a pure substance will only produce one spot on the chromatogram. Mixtures will usually produce more than one spot. In Figure 10.14, Y is a mixture of two substances because there are two spots. By comparing the positions of the spots we can see that Y is a mixture of A and C.

Chromatography works because each substance dissolves in the solvent and will travel at a different speed or rate up the paper.

The chromatography paper is called the stationary phase, and the solvent is called the mobile phase.

The ratio of the distance that the substance moves up the paper compared with the distance that the solvent moves is called the R_f value (retention factor value). The distance of the spot is measured to the leading edge of the spot (the top of the spot), as shown in Figure 10.15.

The formula to find the R_f value is:

$$R_f = \frac{\text{distance moved by substance}}{\text{distance moved by solvent}}$$

The R_f value of a substance is always the same if the same solvent is used. If the solvent changes, the R_f value will change for that substance. R_f values can be used to identify substances.

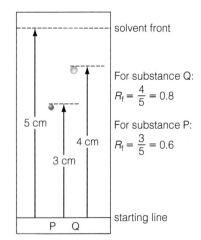

For substance Q:
$R_f = \frac{4}{5} = 0.8$

For substance P:
$R_f = \frac{3}{5} = 0.6$

▲ **Figure 10.15** Finding R_f values

Practice questions

1 Copy and complete Table 10.6 about solids, liquids and gases and their properties. *(4 marks)*

Table 10.6

Property	Solid	Liquid	Gas
Particles arranged closely together	Yes	Yes	No
Particles able to move freely			
Fixed shape			
Fixed volume			
Can be compressed			

2 Copy and complete Figure 10.16 with the changes of state. *(5 marks)*

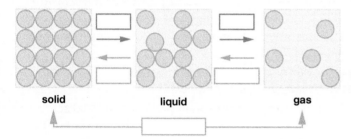

Figure 10.16

3 Table 10.7 gives the melting and boiling points of some substances.

Table 10.7

Substance	Melting point/°C	Boiling point/°C
Water	0	100
Mercury	− 39	360
Helium	− 272	− 269
Chlorine	− 101	− 35
Copper	1083	2567
Carbon	3500	4827

a) Which substance has the highest melting point? *(1 mark)*

b) i) Which substances are liquids at room temperature (25 °C)? *(1 mark)*

ii) Which of the substances in b) i) is liquid over the largest temperature range? *(1 mark)*

c) What needs to be added to all these substances to change them from a solid to a liquid? *(1 mark)*

4 The particle diagram in Figure 10.17 represents a compound. Explain why. *(2 marks)*

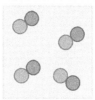

Figure 10.17

5 a) A student wants to separate a mixture of sand and water. Draw a labelled diagram of the apparatus required. *(5 marks)*

b) A student wants to separate salt from salty water. Draw a labelled diagram of the apparatus required. *(5 marks)*

c) Simple distillation can be used to separate two miscible liquids. Describe briefly how this happens. *(4 marks)*

6 Simple chromatography can be used to separate the colours in ink. Figure 10.18 shows a chromatogram with the results for the sample ink (S) and six other inks (1–6).

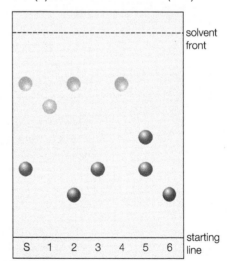

Figure 10.18

a) Describe the procedure that would have been followed to produce this chromatogram. *(4 marks)*

b) Which of the substances 1–6 are pure substances? *(1 mark)*

c) Which substances are present in the sample ink S? *(1 mark)*

d) Calculate the R_f values for each spot in the sample S. Give your answers to one decimal place. *(1 mark)*

11 The Periodic Table, atomic structure and bonding

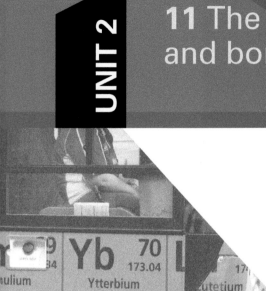

Atomic structure

As seen in Chapter 10, all substances can be classified as elements, compounds or mixtures. They are made up of particles called atoms.

The atoms of any one element are all **identical**. For example, all carbon atoms are the same type of atom, and all magnesium atoms are the same type of atom.

The atoms of different elements are different. For example, an atom of sodium is different from an atom of helium.

Atoms were once thought to be the smallest particles of matter. However, we know now that atoms are made up of three **subatomic** particles, which are protons, electrons and neutrons. Subatomic particles are smaller than the atom (Figure 11.1).

The protons and neutrons are found in the nucleus at the centre of the atom. The electrons are arranged in electron shells orbiting the nucleus.

The mass of an atom depends on the number of these subatomic particles.

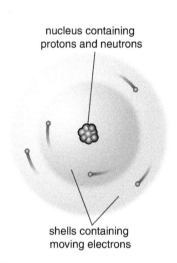

nucleus containing protons and neutrons

shells containing moving electrons

▲ **Figure 11.1** The structure of the atom

Table 11.1 Subatomic particles

Subatomic particle	Relative mass	Relative charge
proton	1	+1
neutron	1	0
electron	1/1840	−1

Protons and neutrons have the same mass. The mass of an electron is very small in comparison.

The presence of the electrons does not affect the mass of an atom by any noticeable amount. So, the mass of an atom depends on the number of protons and neutrons in the nucleus.

An atom is always neutral overall. This is because it has the same number of protons (positive charges) as it does electrons (negative charges).

Every atom of an element has an atomic number. This number is the number of protons in the nucleus of a particular atom. It is used to arrange the elements in the Periodic Table.

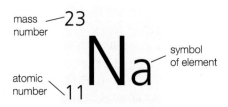

mass number —23

symbol of element

Na

atomic number —11

▲ **Figure 11.2** Element representation from the Periodic Table

Every atom of an element has a mass number, which tells you the total mass of protons and neutrons in the nucleus of that particular atom. The mass number is equal to **the total number of protons and neutrons** in the nucleus of the atom.

The atomic number and mass number of an atom of an element are found in the Periodic Table: an example is shown in Figure 11.2.

The shorthand way of writing the mass number and atomic number with the atom's symbol is $_{11}^{23}$Na.

To work out the numbers of the three subatomic particles in an atom, we need to use the **atomic number** and the **mass number**.

Rule 1 Number of protons = atomic number.

Rule 2 In an atom, number of electrons = number of protons.

This is because an atom is electrically neutral, meaning that it has no charge overall; therefore, it must have the same number of positively charged particles (protons) and negatively charged particles (electrons).

Rule 3 Number of neutrons = mass number – atomic number.

> **Example**
>
> For sodium, $_{11}^{23}$Na, the atomic number is 11 and the mass number is 23.
>
> **Rule 1** Number of protons = atomic number = 11.
>
> **Rule 2** In an atom, number of electrons = number of protons = 11.
>
> **Rule 3** Number of neutrons = mass number – atomic number = 23 – 11 = 12.
>
> So, an atom of sodium has 11 protons, 11 electrons and 12 neutrons.

The arrangement of electrons

Around the nucleus there are several **shells** or orbitals that hold the electrons (Figure 11.3). Electrons have a negative charge. The nucleus has protons in it, so it has a positive charge. The negative electrons are **attracted** to the positive protons in the nucleus. This attraction means that electrons will try to get into a shell as close to the nucleus as they can.

The **maximum** numbers of electrons that each shell can hold in the simple atomic model are given in Table 11.2.

Table 11.2 Numbers of electrons in the atomic shells

Shell	Maximum number of electrons
first	2
second	8
third	8

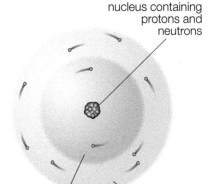

nucleus containing protons and neutrons

first shell, which can have no more than two electrons

second shell, which can have no more than eight electrons

▲ **Figure 11.3** The arrangement of shells around a nucleus

Think about sodium as an example. If an atom has 11 electrons, then two electrons go in the first shell, eight electrons go in the second shell and there is one electron on its own in the third shell.

This is often written as 2,8,1. This is the arrangement of electrons for a sodium atom (Na), and is called its electronic configuration.

Figure 11.4 is a diagram of an atom of sodium, showing the numbers of particles and the arrangement of the electrons. In each shell, electrons will pair up only when they have to. So, in a second shell with five electrons, there will be one pair and three electrons not paired.

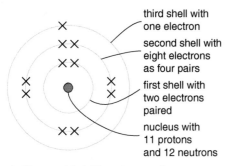

third shell with one electron

second shell with eight electrons as four pairs

first shell with two electrons paired

nucleus with 11 protons and 12 neutrons

▲ **Figure 11.4** The electronic configuration of a sodium atom

For the 2,8,1 configuration, the electrons in the first shell are paired, all the electrons in the second shell are paired and the one electron in the third shell is on its own. Electrons are usually shown as crosses (×).

Figure 11.5 shows the arrangement of the electrons in the atoms of elements with atomic numbers from 1 to 20. When you are drawing a complete atom, you should label the nucleus to show the numbers of protons and neutrons.

The last shell that has electrons is called the **outer shell**. The electrons in the outer shell are **transferred** or **shared** when chemical reactions happen. The movement of these outer shell electrons is the cause of all chemical reactions.

Hydrogen	Helium	Lithium	Beryllium	Boron
1	2	2,1	2,2	2,3

Carbon	Nitrogen	Oxygen	Fluorine	Neon
2,4	2,5	2,6	2,7	2,8

Sodium	Magnesium	Aluminium	Silicon	Phosphorus
2,8,1	2,8,2	2,8,3	2,8,4	2,8,5

Sulfur	Chlorine	Argon	Potassium	Calcium
2,8,6	2,8,7	2,8, 8	2,8,8,1	2,8,8,2

▲ **Figure 11.5** The electronic structures of elements 1 to 20

Test yourself

1 Copy and complete Table 11.3 to show the charges and masses of the subatomic particles.

Table 11.3 Charges and masses of subatomic particles

Subatomic particle	Relative charge	Relative mass
electron	−1	
neutron	0	
proton		1

2 Define the term 'mass number'.
3 a) Draw the structure of an aluminium atom. Include the protons, neutrons and electrons.
   b) What is the electronic configuration of aluminium?
4 Explain why an atom is neutral (has no overall charge).

Show you can **?**

Work out the number of each subatomic particle for the elements in Table 11.4.

Table 11.4 Particles in different elements

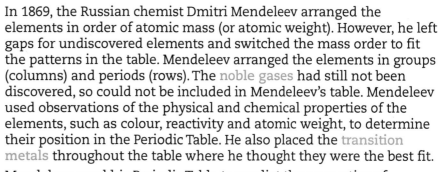

Element	Atomic number	Mass number	Number of protons	Number of neutrons	Number of electrons
helium	2	4			
carbon	6	12			
magnesium	12	24			
hydrogen	1	1			
boron	5	11			

The Periodic Table

The **Periodic Table** lists all known elements. An element cannot be broken down into anything simpler by chemical reactions.

History of the Periodic Table

▲ **Figure 11.6** The Russian scientist Dmitri Mendeleev

In 1869, the Russian chemist Dmitri Mendeleev arranged the elements in order of atomic mass (or atomic weight). However, he left gaps for undiscovered elements and switched the mass order to fit the patterns in the table. Mendeleev arranged the elements in groups (columns) and periods (rows). The noble gases had still not been discovered, so could not be included in Mendeleev's table. Mendeleev used observations of the physical and chemical properties of the elements, such as colour, reactivity and atomic weight, to determine their position in the Periodic Table. He also placed the transition metals throughout the table where he thought they were the best fit.

Mendeleev used his Periodic Table to predict the properties of elements that had not yet been discovered. These elements had almost exactly these properties when they were discovered.

Over time, scientific ideas change as more evidence is gathered. The modern Periodic Table is arranged in order of atomic number and contains more elements than Mendeleev's table (all the gaps have been filled). Noble gases (Group 0) have been added – these had not been discovered in Mendeleev's time because they are so unreactive. The modern Periodic Table also has a block for the transition metals.

The **periods** are the horizontal rows of the Periodic Table. (The first period only contains two elements – hydrogen and helium.)

The **groups** are the vertical columns of the Periodic Table.

H

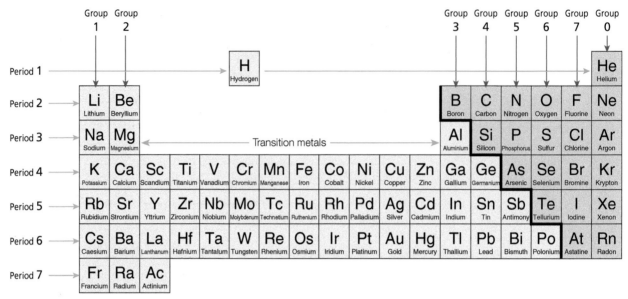

▲ **Figure 11.7** The Periodic Table

Figure 11.7 shows the symbols used as short-hand to stand for the elements and their common names.

The thick black line on the Periodic Table divides metals, which are listed on the left of the line, from **non-metals**, on the right. (It is shown on the Periodic Table in the Data Leaflet provided in the examination.) The metallic character of the elements decreases as you move from left to right across the Periodic Table. Silicon and germanium are called **semi-metals** as they show properties of both metals and non-metals.

In the Periodic Table, elements that have similar properties are grouped together. This means that elements that react in a similar way and are similar physically are in the same vertical column or group. Each of these groups of elements has a number, like those shown above, but some of the groups also have names.

▶ **Group 1** is a group of reactive metals, also called the alkali metals.

▶ **Group 2** is a group of metals, also called the alkaline earth metals.

▶ **Group 7** is a group of reactive non-metals, also called the halogens.

▶ **Group 0** (also called **Group 8**) is a group of non-reactive (inert) non-metals, also called the noble gases, which are helium, neon, argon, krypton, xenon and radon.

▶ The block of elements between Group 2 and Group 3 is called the transition metals.

Of all the known elements, 11 are gases, two are liquids and the rest are solids.

▶ The 11 gases are hydrogen, nitrogen, oxygen, fluorine, chlorine and the six noble gases.

▶ The two liquids are bromine (a non-metal) and mercury (a metal).

Some non-metallic elements exist as diatomic molecules. This means that two identical atoms are joined together. When you write the formulae for these elements, they should be written with a subscript 2 after the symbol, for example H_2 and O_2.

Tip

Make sure that you learn the names of the groups carefully: Group 1 is the alkali metals – alkaline metals will not be accepted.

There are **seven** diatomic elements to remember:

Hydrogen, H_2 Nitrogen, N_2 Oxygen, O_2 Fluorine, F_2

Chlorine, Cl_2 Bromine, Br_2 Iodine, I_2

The formula for sodium is Na but the formula for hydrogen is H_2. The formula for magnesium is Mg but the formula for fluorine is F_2.

Compounds

A compound is a substance in which two or more elements are chemically joined together. Compounds are formed when electrons move from one atom to another or when two atoms share electrons between them. It is the electrons in the outer shell that move or are shared.

Many compounds have already been named in this book. Chemists do not write the name of a compound when they can use a chemical short-hand. The symbols of the elements can be used to write **chemical formulae**. Simple compounds are formed from two elements. They can be formed from a metal and a non-metal or from two non-metals.

Simple compounds containing a metal and a non-metal

▶ Sodium chloride is a simple compound of sodium and chlorine.

▶ Magnesium oxide is a simple compound of magnesium and oxygen.

▶ Copper(II) sulfide is a simple compound of copper and sulfur.

Sodium chloride

The compound sodium chloride has one sodium atom (Na) and one chlorine atom (Cl). Chemists write these together as NaCl. Compounds of Group 1 elements combined with Group 7 elements always have formulae like this, where there is one atom of each element. For example, potassium bromide is KBr.

Potassium oxide

The compound potassium oxide has two potassium atoms (K) and one oxygen atom (O). Chemists write these together as K_2O. The subscript 2 after the potassium means that there are two atoms of this element. Compounds of Group 1 elements combined with Group 6 elements always have formulae like this, where two atoms of the Group 1 element combine with one atom of the Group 6 element. For example, sodium sulfide is Na_2S.

Magnesium chloride

The compound magnesium chloride has one magnesium atom (Mg) and two chlorine atoms (Cl). Chemists write these together as $MgCl_2$. Compounds of Group 2 elements combined with Group 7 elements always have formulae like this, where one atom of the Group 2 element is combined with two atoms of the Group 7 element. For example, calcium fluoride is CaF_2.

Magnesium oxide

The compound magnesium oxide has one magnesium atom (Mg) and one oxygen atom (O). Chemists write magnesium oxide as MgO.

Compounds of Group 2 elements combined with Group 6 elements always have formulae like this, where one atom of the Group 2 element is combined with one atom of the Group 6 element. For example, calcium oxide is CaO.

Aluminium chloride

The compound aluminium chloride has one aluminium atom (Al) and three chlorine atoms (Cl). Chemists write aluminium chloride as $AlCl_3$. Compounds of Group 3 elements combined with Group 7 elements always have formulae like this, where one atom of the Group 3 element is combined with three atoms of the Group 7 element. For example, aluminium fluoride is AlF_3.

Aluminium oxide

The compound aluminium oxide has two aluminium atoms (Al) and three oxygen atoms (O). Chemists write aluminium oxide as Al_2O_3. Compounds of Group 3 elements combined with Group 6 elements always have formulae like this, where two atoms of the Group 3 element are combined with three atoms of the Group 6 element. For example, aluminium sulfide is Al_2S_3.

Simple transition metal compounds

Most transition metal elements form compounds with similar chemical formulae to the Group 2 elements. (Exceptions are silver, which forms compounds like the Group 1 elements, and iron(III), which forms compounds like Group 3 elements.) The (II) in copper(II) sulfide tells you that copper is acting like a Group 2 element when it is forming compounds. Copper(II) sulfide is CuS (like magnesium oxide). Iron(III) oxide is similar to aluminium oxide, so its formula is Fe_2O_3. If there is no roman numeral in brackets after the name of the transition metal element, assume it is II (except silver, which is I). For example, zinc oxide is ZnO (like MgO), copper chloride is $CuCl_2$ (like $MgCl_2$) and silver chloride is AgCl (like NaCl).

Simple compounds containing two non-metals

In a simple compound that contains two non-metals, the element that is further right in the Periodic Table (ignoring Group 0) and/or higher up changes its name to end in –**ide**.

For example:

▶ carbon dioxide is a simple compound formed from carbon and oxygen

▶ hydrogen chloride is a simple compound formed from hydrogen and chlorine.

Formulae of more complex compounds

The back of the Data Leaflet shows some molecular ions that may be used to write formulae. The most commonly used ones are carbonate, hydrogencarbonate, hydroxide, nitrate and sulfate. Carbonates and sulfates form compounds with similar numbers of atoms in the formulae. Hydrogencarbonates, hydroxides and nitrates again form compounds with similar numbers of atoms in the formulae.

Carbonates and sulfates

- Group 1 elements form carbonates that have formulae like Na_2CO_3 (sodium carbonate) and K_2CO_3 (potassium carbonate). The formulae of Group 1 sulfates are similar to this. Sodium sulfate is Na_2SO_4 and potassium sulfate is K_2SO_4.
- Group 2 elements form carbonates that have formulae like $MgCO_3$ (magnesium carbonate) and $CaCO_3$ (calcium carbonate). Again, the formulae of Group 2 sulfates are similar to this. Magnesium sulfate is $MgSO_4$ and calcium sulfate is $CaSO_4$.
- Aluminium sulfate is $Al_2(SO_4)_3$, and iron(III) sulfate is $Fe_2(SO_4)_3$.

Hydrogencarbonates, hydroxides and nitrates

- Group 1 elements form hydrogencarbonates that have formulae like $NaHCO_3$ (sodium hydrogencarbonate) and $KHCO_3$ (potassium hydrogencarbonate).
- Group 1 elements form hydroxides that have formulae like $NaOH$ (sodium hydroxide) and KOH (potassium hydroxide).
- Group 1 elements form nitrates that have formulae like $NaNO_3$ (sodium nitrate) and KNO_3 (potassium nitrate).
- Group 2 elements form hydrogencarbonates that have formulae like $Mg(HCO_3)_2$ (magnesium hydrogencarbonate) and $Ca(HCO_3)_2$ (calcium hydrogencarbonate).
- Group 2 elements form hydroxides that have formulae like $Mg(OH)_2$ (magnesium hydroxide) and $Ca(OH)_2$ (calcium hydroxide).
- Group 2 elements form nitrates that have formulae like $Mg(NO_3)_2$ (magnesium nitrate) and $Ca(NO_3)_2$ (calcium nitrate).
- Aluminium hydroxide is $Al(OH)_3$, and aluminium nitrate is $Al(NO_3)_3$.

More complex compounds of transition metals

Again, the (II) in copper(II) sulfate tells you that copper is acting like a Group 2 element when it is forming compounds. Copper(II) sulfate is $CuSO_4$ (like magnesium sulfate). Iron(III) nitrate is similar to aluminium nitrate, so its formula is $Fe(NO_3)_3$. If there is no roman numeral in brackets after the name of the transition metal element, assume it is II (except silver, which is I). For example, zinc hydroxide is $Zn(OH)_2$ (like $Mg(OH)_2$), copper carbonate is $CuCO_3$ (like $CaCO_3$) and silver nitrate is $AgNO_3$ (like $NaNO_3$).

Naming compounds

Compounds can be named from their formulae. Compounds usually have two parts to their name.

- Al_2O_3 contains two Al (aluminium) and three O (oxygen). Oxygen is further up the Periodic Table and further to the right than aluminium. Therefore, its name changes from oxygen to oxide – which becomes the second part of the name. The name of this compound is **aluminium oxide**.

Tip

When writing formulae, it is important to write the symbols of the elements correctly.

▶ All elements have one capital letter, for example carbon is C and sulfur is S.

▶ If there are two letters in the symbol, the first will be a capital and the second will be lower case, for example copper is Cu and magnesium is Mg.

This means that to count the number of elements in a compound you can simply count the number of capital letters. For example, $CaSO_4$ has three capital letters, so it has three elements.

▶ CaF_2 contains one Ca (calcium) and two F (fluorine). Fluorine is further up the Periodic Table and further to the right than calcium. Therefore, its name changes from fluorine to fluoride – which becomes the second part of the name. The name of this compound is **calcium fluoride**.

▶ K_2SO_4 contains two K (potassium) and one SO_4 (sulfate – from the back of the Data Leaflet). The name of this compound is **potassium sulfate**.

Some compounds, such as water (H_2O) and methane (CH_4), have only one name. It is important to learn the formulae of these compounds. Other compounds with special names are hydrochloric acid (HCl) and sulfuric acid (H_2SO_4).

Atoms in formulae

The formula of a substance can tell us the number of atoms of each element in the compound, and we can also use the formula to work out the total number of atoms present.

Example

The formula of sulfuric acid is H_2SO_4.

So, in total there are:

- 3 elements present (hydrogen, sulfur and oxygen)
- 7 atoms present (2 H + 1 S + 4 O).

Some formulae may contain brackets: for example, calcium hydroxide has the formula $Ca(OH)_2$.

So, in total there are:

- 3 elements present (calcium, oxygen and hydrogen)
- 5 atoms present (1 Ca + 2 O + 2 H).

2 H atoms 1 S atom 4 O atoms

▲ **Figure 11.8** Sulfuric acid formula

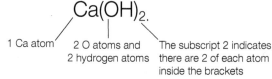

1 Ca atom 2 O atoms and 2 hydrogen atoms The subscript 2 indicates there are 2 of each atom inside the brackets

▲ **Figure 11.9** Calcium hydroxide formula

Show you can

What is the total number of elements and atoms in each of the following compounds?

1 $CaCO_3$
2 $Mg(OH)_2$
3 $Al_2(SO_4)_3$

Test yourself

5 Copy and complete the following sentences about the first Periodic Table.
In 1869, the Russian chemist Dmitri _____ arranged the elements in order of atomic _____. However, he left gaps for undiscovered elements and switched the mass order to fit the patterns in the table. The _____ _____ had still not been discovered so could not be included in Mendeleev's table. Mendeleev used his Periodic Table to predict the properties of elements that had not yet been discovered.

6 The modern Periodic Table is ordered by atomic number and is organised in rows and columns.
a) What name is given to:
i) a row in the Periodic Table?
ii) a column of the Periodic Table?

b) Copy and complete Table 11.5 listing the group names in the Periodic Table.

Table 11.5 Groups in the Periodic Table

Group	Name
1	
2	
	halogens
	noble gases

c) Hydrogen and oxygen are both elements in the Periodic Table. They are described as diatomic molecules. What is meant by the term 'diatomic'?

7 Copy and complete Table 11.6 about compounds.

Table 11.6 Compounds

Compound	Formula	Number of elements present
sodium chloride	NaCl	2
magnesium oxide	MgO	
	$AlCl_3$	2
	$Ca(OH)_2$	

The alkali metals

Appearance

All the alkali metals are soft, grey solids. They are all very reactive metals and are stored in oil to stop them reacting with the moisture in the air. They can be cut easily with a knife or scalpel. When they are cut they show a shiny surface that starts to tarnish (go dull) as the metal reacts with the moisture in the air. They must not be allowed to touch human skin, as it too contains moisture.

Reaction with water

All the alkali metals react with water. If a piece of lithium is added to water it floats on the surface, moves around, and bubbles of gas are produced. Heat is released and a colourless solution remains.

If a small piece of sodium is added to water, it floats on the surface and moves about. There are bubbles as a gas is produced. It sometimes burns with a yellow flame, and will eventually disappear. The reaction is very vigorous.

If a small piece of potassium is added to water, it floats on the water and moves about on the surface. There are bubbles as a gas is produced. It burns with a lilac flame, and will eventually disappear, with a crack. The reaction is even more vigorous than the reaction of sodium with water. Potassium is more reactive than sodium.

As you move down Group 1, the elements become more reactive. Rubidium and caesium react violently with water. The lower down the Periodic Table a Group 1 element is, the more reactive it is, and so the faster and more vigorously it will react with water.

Tip

A common exam question is to describe the observations when an alkali metal is placed in water. The observations of bubbles or hisses or fizzes all indicate that a gas is given off, so will only be awarded 1 mark if listed together.

When Group 1 elements react with water, they form a colourless solution (metal hydroxide) and release a gas, which is hydrogen. The word equations for the reactions are:

$$lithium + water \rightarrow lithium\ hydroxide + hydrogen$$
$$sodium + water \rightarrow sodium\ hydroxide + hydrogen$$
$$potassium + water \rightarrow potassium\ hydroxide + hydrogen$$

All Group 1 metals react with water in a similar way. The balanced symbol equations for the reactions are always balanced using 2 before the metal, 2 before the water and 2 before the metal hydroxide.

The balanced symbol equations for these reactions are:

$$2Li + 2H_2O \rightarrow 2LiOH + H_2$$
$$2Na + 2H_2O \rightarrow 2NaOH + H_2$$
$$2K + 2H_2O \rightarrow 2KOH + H_2$$

The alkali metals all react in the same way because they all have one electron in their outer shell.

Group 0 (the noble gases)

Group 0 (or 8) of the Periodic Table contains only gases. These are the noble gases. They are all colourless, and they are called noble as they do not react very well with any other elements. Because they are so unreactive, the noble gases were discovered much later than many other elements. Although they are found in air, separating them from the air was very difficult. All of the noble gases except helium were discovered in the late 19th century by Lord Rayleigh. Helium, neon and argon do not form any compounds. They are chemically inert.

The reason why all the noble gases are unreactive is because they all have a full outer shell of electrons, making them stable, as shown in Figure 11.10.

Helium atom

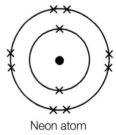
Neon atom

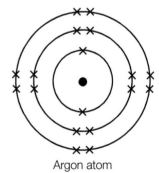
Argon atom

⌃ **Figure 11.10** Electronic configuration of the first three noble gases

Arrangement of electrons from the Periodic Table

You can use the Periodic Table to work out the arrangement of the electrons in an atom, or vice versa.

The group number gives the number of electrons in the outer shell.

The period number gives the number of shells, so the outer shell number will be the period number.

Example

Magnesium (Mg) is in Group 2 and in Period 3:

- Period 3 → three shells are being used.
- Group 2 → two electrons in the outer (third) shell.
- The first and second shells must be full, with two and then eight electrons.
- The arrangement of electrons for magnesium is 2,8,2.

Oxygen (O) is in Group 6 and in Period 2:

- Period 2 → two shells are being used.
- Group 6 → six electrons in the outer (second) shell.
- The first shell must be full, with two electrons.
- The arrangement of electrons for oxygen is 2,6.

Potassium (K) has an electronic arrangement of 2,8,8,1:

- Four shells are being used → Period 4.
- One electron is in the outer (fourth) shell → Group 1.

Test yourself

8 Group 1 elements are the alkali metals.
   a) Name two alkali metals.
   b) Describe the appearance of an alkali metal.
   c) Why are alkali metals stored in oil?
9 When potassium is added to water there is a violent reaction.
   a) Give three observations when potassium is added to water.
   b) State two safety precautions required during this experiment.
   c) i) Write a word equation for the reaction of potassium with water.
      ii) Write a balanced symbol equation for the reaction of potassium with water.
   d) Give two ways in which the reaction of sodium with water would be different to potassium reacting with water.
10 a) What group of the Periodic Table do the noble gases belong to?
   b) Suggest why the noble gases were not included in Mendeleev's Periodic Table.
   c) Draw the electronic structure of neon.

Show you can

Copy and complete Table 11.7 giving information about different elements.

Table 11.7 Elements

Electronic arrangement	Group number	Period number
2,8,3		
2,5		
	1	3

Formation of compounds

Compounds formed from a metal and a non-metal are called ionic compounds. In ionic compounds, electrons are given away by one atom, and these electrons are taken by another atom. Examples of ionic compounds include sodium chloride and magnesium oxide. Compounds that only contain non-metals are called covalent compounds. The atoms in covalent compounds share electrons. Examples of covalent compounds include hydrogen chloride and water. Atoms of elements always react to achieve a full outer shell of electrons (like the noble gases) and in doing so they become more stable.

Ionic compounds

Ionic compounds are compounds that contain a metal and a non-metal. Examples are sodium chloride (NaCl) and magnesium oxide (MgO).

Ionic compounds are made up of **ions**, which are charged particles. An ionic compound contains a positive ion and a negative ion.

Simple ions are those that are formed when an atom of an element gains or loses electrons, for example Na^+, O^{2-}, Mg^{2+} and Cl^-.

Formation of ions from atoms

When an ionic compound forms from the atoms of its elements, a transfer of electrons occurs. Metal atoms lose electrons and give them to non-metal atoms, which gain electrons. Each will lose or gain enough electrons to give it a full outer shell and make it more stable.

When a metal atom loses electrons, it becomes a positively charged ion, for example sodium (Na) loses one electron to form a sodium ion (Na^+).

When a non-metal atom gains electrons, it becomes a negatively charged ion, for example chlorine (Cl) gains one electron to form a chloride ion (Cl^-).

There are two main combinations used to make an ionic compound:

▶ Group 1 metal with Group 7 non-metal, e.g. NaCl
▶ Group 2 metal with Group 6 non-metal, e.g. MgO

Example

1 Sodium chloride

A sodium atom has an electronic configuration of 2,8,1.

When it reacts with a chlorine atom (electronic configuration 2,8,7), the single outer electron of the sodium atom is given to the outer shell of the chlorine atom.

Sodium now has only 10 electrons (2,8), but it still has 11 protons in its nucleus, so it has a charge of '+1'.

The sodium atom is written as Na. The sodium ion is written Na^+.

Simple positive ions have the same name as the atom, so it is called a sodium ion.

Chlorine now has 18 electrons (2,8,8), but it still has 17 protons in its nucleus, so it has a charge of '–1'.

The chlorine atom is written as Cl. The chloride ion is written as Cl^-.

Simple negative ions change the end of their name to **–ide**, so it is called a chloride ion.

The sodium and chloride ions are attracted to each other and form the ionic compound.

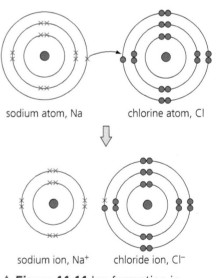

sodium atom, Na   chlorine atom, Cl

sodium ion, Na^+ chloride ion, Cl^-

▲ **Figure 11.11** Ion formation in sodium chloride

(An ionic bond is a strong electrostatic force of attraction between oppositely charged ions in an ionic compound. Ionic bonds are strong and require a large amount of energy to break them.)

Figure 11.11 summarises the process of ion formation in sodium chloride.

The compound formed is called sodium chloride (as it contains sodium ions and chloride ions).

2 Magnesium oxide

Only one magnesium atom is required for each oxygen atom when magnesium oxide forms, as each magnesium atom loses two electrons and each oxygen atom gains two electrons (see Figure 11.12).

A magnesium ion is Mg^{2+} as it has 10 electrons (2,8) but 12 protons (the atomic number is 12).

An oxide ion is O^{2-} as it has 10 electrons (2,8) but eight protons (the atomic number is 8).

The formula of magnesium oxide is MgO.

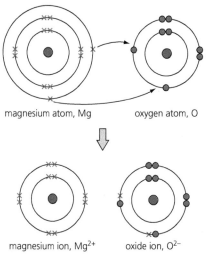

magnesium atom, Mg   oxygen atom, O

magnesium ion, Mg^{2+} oxide ion, O^{2-}

▲ **Figure 11.12** Ion formation in magnesium oxide

Test yourself

11 Give two differences between ionic and covalent compounds.
12 Copy and complete the following sentences about ionic bonding in magnesium oxide.
   Magnesium is a metal in Group 2 of the Periodic Table. To form an ion it must _____ two electrons. Oxygen is a _____ in Group 6 of the Periodic Table. It must gain _____ electrons to form an ion. The formula of a magnesium ion is Mg^{2+} and the formula of an oxide ion is _____.
   In the compound _____ _____ the magnesium ions and the oxide ions are held together by strong forces of attraction between the oppositely charged ions.

Show you can

Use diagrams of electronic arrangements of both atoms and ions to show how:

1 lithium fluoride forms
2 calcium oxide forms.

Covalent compounds

When non-metal atoms bond to form compounds they share electrons. Non-metal atoms share electrons to get a full outer shell. The shared electrons count towards the outer shell electrons for both atoms. The shared electrons make a covalent bond between the atoms, which holds them together. This bond is strong and requires a large amount of energy to break.

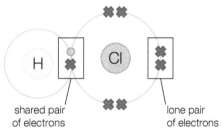

shared pair
of electrons

▲ **Figure 11.13** Bonding in hydrogen

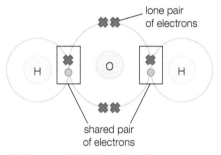

shared pair   lone pair
of electrons   of electrons

▲ **Figure 11.14** Bonding in hydrogen chloride

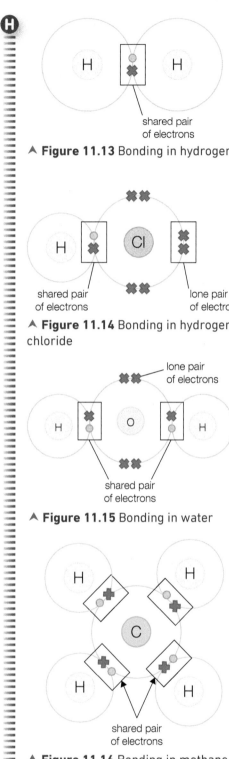

lone pair
of electrons

shared pair
of electrons

▲ **Figure 11.15** Bonding in water

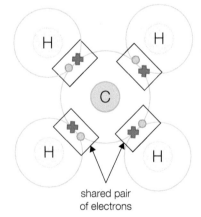

shared pair
of electrons

▲ **Figure 11.16** Bonding in methane

A **hydrogen** molecule (H_2) is diatomic, which means there are two hydrogen atoms making up the molecule. Each hydrogen atom contributes one electron for sharing, which means that each hydrogen has a full outer shell of two electrons (Figure 11.13).

A **hydrogen** molecule can be represented using a 'stick diagram': each line in the diagram represents one shared pair of electrons (a single covalent bond):

H—H

Hydrogen chloride is a molecule made from one atom of hydrogen and one atom of chlorine. Each atom contributes one electron for sharing. This means that each atom has a full outer shell. The pairs of electrons not involved in bonding are described as lone pairs of electrons. Hydrogen chloride has 3 lone pairs of electrons (Figure 11.14).

A hydrogen chloride molecule can be represented using a 'stick diagram': each line in the diagram represents one shared pair of electrons (a single covalent bond):

H—Cl

Water is a molecule made from one atom of oxygen and two atoms of hydrogen. Each oxygen atom contributes one electron for sharing with each hydrogen atom. This means that every atom has a full outer shell. This means that the oxygen contributes a total of two electrons for sharing. The pairs of electrons not involved in bonding are described as lone pairs of electrons. Water has two lone pairs of electrons (Figure 11.15).

A water molecule can be represented using a 'stick diagram': each line in the diagram represents one shared pair of electrons (a single covalent bond):

O
H   H

Methane is a molecule made from one atom of carbon and four atoms of hydrogen. Each carbon atom contributes one electron for sharing with each hydrogen atom. This means that every atom has a full outer shell. This also means that the carbon contributes a total of four electrons for sharing. Methane has no lone pairs of electrons (Figure 11.16).

A methane molecule can be represented using a 'stick diagram'. Each line in the diagram represents one shared pair of electrons (a single covalent bond):

H
|
H—C—H
|
H

Practice questions

1 a) Copy and complete Table 11.8 about the number of subatomic particles in different elements. *(5 marks)*

Table 11.8

Element	Atomic number	Mass number	Number of protons	Number of electrons	Number of neutrons
fluorine	9	19			
nitrogen			7	7	7
calcium	20	40			
potassium		39			20
		40	18		

b) Elements and compounds can be represented by chemical symbols, but they must be written correctly. Write down the correct formula for the compounds copper sulfate and magnesium chloride. (You may find your Data Leaflet helpful.)
  i) Copper sulfate
  Choose from $CuSO_4$, CoSu, $CuSO_2$, $CUSO_4$. *(1 mark)*
  ii) Magnesium chloride
  Choose from MGCl, MC, $MgCl_2$. *(1 mark)*

2 Figure 11.17 shows an atom of beryllium.

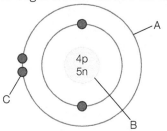

4p
5n

Figure 11.17

a) Name the parts labelled A, B and C on the diagram. *(3 marks)*
b) A silicon atom has 14 electrons. Draw a diagram to show how all the electrons are arranged in an atom of silicon. *(1 mark)*
c) What is meant by the term 'atomic number'? *(1 mark)*

3 Table 11.9 contains information about the structure of four elements: W, X, Y and Z. (You may find your Data Leaflet helpful.)

Table 11.9

Element	Number of protons	Number of neutrons	Number of electrons
W	8	8	8
X	6	6	6
Y	15	16	15
Z	11	12	11

a) Calculate the mass number of element Y. *(1 mark)*
b) Name the element X. *(1 mark)*
c) Which element (W, X, Y or Z) has six electrons in its outer shell? *(1 mark)*
d) Which element (W, X, Y or Z) is an alkali metal? *(1 mark)*

4 a) Copy and complete the diagrams below to show the arrangement of **all** of the electrons in a magnesium atom and in an oxygen atom.
  i) Magnesium atom *(1 mark)*

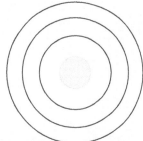

  ii) Oxygen atom *(1 mark)*

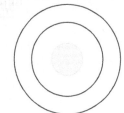

b) Describe, in terms of electrons and ions, how the compound magnesium oxide is formed from an atom of magnesium and an atom of oxygen. *(4 marks)*

5 The Periodic Table has been developed over a number of years. Dmitri Mendeleev was one of the chemists who helped to do this.
a) In what order did Mendeleev arrange the elements? *(1 mark)*
b) Why did he leave gaps between some of the elements? *(1 mark)*
c) Give two differences between Mendeleev's table and the modern Periodic Table. *(2 marks)*

6 Water has the formula H_2O.
a) Draw a diagram of the bonding in a molecule of water showing all the electrons. *(3 marks)*
b) On the diagram in a), label:
  i) a shared pair of electrons *(1 mark)*
  ii) a lone pair of electrons. *(1 mark)*
c) What name is given to the type of bonding in water? *(1 mark)*

12 Materials

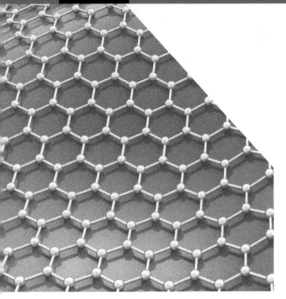

Specification points

This chapter covers sections 2.5.1 to 2.5.14 of the specification. It is about the properties of materials, different types of materials including emergent materials, and also the use of materials to fight crime.

Materials

Natural and synthetic materials

A material is any substance that can be used to make something. For example, wood is a material that can be used to make a chair, cotton is a material that can be used to make shirts, and copper is a material that can be used to make electrical wiring.

The raw materials that are used to make everything come from five different sources:

▶ the Earth, for example iron, limestone, granite and aluminium
▶ the sea, for example salt and bromine
▶ the air, for example oxygen, argon and helium
▶ crude oil, for example plastics and drugs
▶ living things (plants and animals), for example wool, cotton, wood and silk.

The raw materials may need to be processed before they are useful. Sometimes this involves some sort of chemical process.

Materials that do not need to be processed by chemical methods are called natural materials.

▶ Natural materials include wood, wool, salt, cotton, limestone and granite.

If a chemical reaction is needed to make the useful material, it is called a synthetic material. Synthetic materials are also called man-made materials.

▶ Synthetic materials include iron, aluminium, glass and plastics.

Properties of materials

What a material is used to make is dictated by its properties:

▶ a windscreen made out of aluminium would be useless
▶ electrical wires made out of nylon would not work.

The material must have the right properties for the job it is doing. There are many different properties that may need to be considered – some are shown in Figure 12.1.

Scientists can examine the properties of materials. They can then make a decision about which material is best for a particular job.

Tip

A common examination question on properties of materials will ask you to analyse the properties of a list of materials and make a choice as to the most suitable one for a particular job. It is important to choose the best material and then only list the relevant or most important properties. Simply listing all the properties from the question usually cannot gain full marks.

Physical properties (what the material is like)

- Melting and boiling points
- Strength
- Hardness
- Density
- Ability to conduct electricity
- Appearance (is it shiny, will its colour fade in the Sun, can it be coloured easily?)

Chemical properties (how the material reacts)

- With water
- With air
- With other chemicals
- Perhaps with items of food if the material is to be used in a kitchen

Cost

- This is very important because some materials are very expensive and some are much cheaper, which may affect the overall cost of the item.
- Some raw materials may be expensive or the material itself may be expensive to manufacture.
- Sometimes cost is irrelevant, such as producing the fastest Formula One racing car.

Availability (a measure of how easy it would be to get the material)

- Is it available in this country?
- Is there enough of it?

▲ **Figure 12.1** Properties of materials

The main types of materials we use are shown in Table 12.1.

Table 12.1 The main types of materials and their uses

Type	Metals	Glass	Ceramics	Fibres	Plastics
Examples	iron, aluminium, copper, gold, silver	soda glass, pyrex (heat-resistant glass)	china, bricks, pottery	nylon, cotton, silk	polythene, polypropene, PVC
Uses	electrical wiring, saucepans, building, jewellery	windscreens, vases, bottles for chemicals	plates, cups, house bricks	clothing, curtains	plastic bags, plastic bottles, window frames
Properties	good conductors of electricity and heat, strong, hard, malleable (can be hammered into shape), ductile (can be drawn out into wires), high melting point, lustrous (shiny), some are reactive	brittle, transparent, hard, can be shaped on heating, does not conduct electricity, unreactive	brittle, hard, high melting point, unreactive, do not conduct electricity	flexible, low density, can be woven, do not conduct electricity	flexible, can be moulded on heating, do not conduct electricity, low density

Decline of traditional materials

Traditional materials such as linen are not used as much nowadays because cheaper materials with better properties are now available. Linen had been used to make tablecloths, napkins and more formal clothing such as men's shirts and women's petticoats.

The decline of the linen industry in Northern Ireland began in the early 1900s, when newer materials became available and also when people started to dress less formally. Paper napkins and handkerchiefs replaced linen ones.

Cotton, nylon and blends of other materials have replaced linen materials. These newer materials are easier to process, making them cheaper, and they crease less easily and are easier to iron.

There is a lot of research being carried out in the field of material science to find better or more suitable materials for the ever-evolving range of products we have.

Smart materials

A smart material is a material that changes how it behaves (its properties) as a result of a change in its surroundings, for example, heat or light exposure.

Thermochromic materials change colour due to a change in temperature.

Some paints and dyes can change colour when heated. They are used in mood rings and in t-shirts that change colour when your body heats them up. Thermochromic dyes can also be added to plastics to make baby feeding spoons or bottles – these will change colour if the food is too hot. Forehead thermometers change colour depending on the temperature of the skin (Figure 12.2).

Photochromic materials change colour depending on their exposure to light.

Some paints and dyes can change colour when exposed to light. They are used in lenses for glasses – these darken in colour when it is sunny and then can act as sunglasses (Figure 12.3).

▲ **Figure 12.2** A thermochromic strip in a forehead thermometer

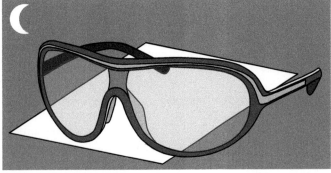

▲ **Figure 12.3** Photochromic dyes can be used in glasses

Nanomaterials

Something that is described as **nano** is very small. A measure of 1 millionth of a millimetre or 0.000000001 m (usually written 1×10^{-9} m) is called a nanometre. Something the size of a nanometre cannot be seen with the naked eye. The sizes of atoms are measured in nanometres. A nanomaterial is one that contains only a few hundred atoms.

Nanotechnology involves the design of **nanomaterials** that are very small – often only a few nanometres across. Nanomaterials are more sensitive to light, heat, electrical conductivity and magnetism. These very small particles have a large surface area, which gives them different properties compared with traditional materials.

Uses of nanoparticles

Sun creams

The latest sun-protection creams use nanotechnology. Older sun-protection creams left a white residue of zinc oxide. Newer creams, using nanoparticles of zinc oxide or titanium dioxide, give better skin coverage and more protection from the Sun's ultraviolet rays. They also rub on as a clear film.

Sterilising

Nanoparticles of silver nitrate are used in wound dressings in hospitals. Silver nanoparticles kill bacteria and prevent wounds from becoming infected, which helps healing.

Silver nanoparticles are also used in sterilising sprays in the beauty industry and in medicine to sterilise equipment.

The safety of nanoparticles

Nanoparticles are relatively new to material science and so, although thorough research will have been done by scientists, not all the risks may be evident yet.

One way in which nanoparticles may be harmful is that some may be able to enter human cells by penetrating their cell membranes and so cause damage to the body's cells. As nanomaterials become more widely used there may also be issues with disposing of them safely.

Graphene

Material scientists are constantly researching the possibilities of new materials. A carbon-based material called **graphene** (Figure 12.4) was first isolated in 2004 by scientists in Manchester, who were awarded a Nobel Prize in 2010 for their work.

Graphene was the world's first two-dimensional material and is 1 million times smaller than the diameter of a single human hair. It is a layer of graphite that is only one atom thick. Graphene is 200 times stronger than steel, but incredibly lightweight and flexible. It is a good conductor of electricity and heat. It is also transparent.

At first the uses were not evident, but as more research was carried out it was discovered that graphene had many applications in the medicine, electronics, energy and transport industries.

Graphene can increase the lifespan of a battery – this means that devices can hold their power for longer and be charged more quickly. The batteries could also be more flexible and lighter, which means that they could be sewn into clothing.

Storage of solar power may also be made possible by using graphene in what would be essentially giant packs of batteries.

▲ **Figure 12.4** A sheet of graphene

Emergent materials

Alongside graphene, there has been much research into other carbon-based materials. Buckminsterfullerene was identified in the 1980s. This molecule consists of 60 carbon atoms (so it has the formula C_{60}), and its shape resembles a football, so it is often called a 'buckyball'. Other similar molecules such as C_{70}, C_{76} and C_{84} have been produced: molecules in this family are called fullerenes. Scientists at the University of Sussex were awarded a Nobel Prize in 1996 for their work on fullerenes.

Fullerenes are being used:

▶ for delivery of drugs to specific parts of the body or a specific cell – the drug is often carried in the hollow centre of a fullerene molecule

▶ as catalysts – due to their large surface area.

▲ **Figure 12.5** C_{60} has a shape that resembles a football

Scientists are also investigating other similar buckyball structures made from boron or silicon rather than carbon.

Carbon nanotubes are cylindrical fullerenes, sometimes called buckytubes. They can be thought of as tubes of graphene sheets. Carbon nanotubes have a very high tensile strength and are good electrical and heat conductors. This makes them suitable materials for many uses, for example to reinforce sports equipment such as tennis racquets and golf clubs while keeping them lightweight.

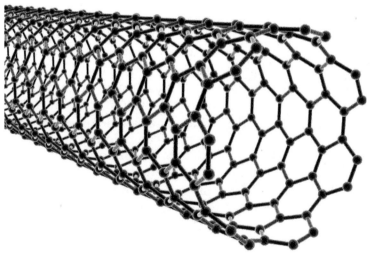

▲ **Figure 12.6** A carbon nanotube

Test yourself ✎

1 Materials can be described as natural or synthetic.
   a) What is meant by the term 'natural material'?
   b) Give two examples of natural materials.
2 Scientists are investigating uses for newly developed materials such as smart materials.
   a) What is a smart material?
   b) Thermochromic materials are one type of smart material.
      i) What is meant by the term 'thermochromic'?
      ii) State and explain one use of a thermochromic material.
3 Nanotechnology involves the use of very small particles called nanoparticles.
   a) What size is a nanoparticle?
   b) Give one use of nanoparticles.
   c) Explain why some people may be reluctant to use nanoparticles.
4 Graphene was the world's first two-dimensional material.
   a) What element is graphene made from?
   b) State one use of graphene.

Show you can

Table 12.2 gives information about some plastics.

Table 12.2

Plastic	Properties	Colours available	Cost
PVC	hard, keeps it shape, weather resistant	wide range of colours	medium
nylon	hard, tough, long lasting	white or cream	high
polythene	soft, flexible, good electrical insulator	wide range of colours (fades easily)	medium
plasticised PVC	soft, flexible, good electrical insulator	wide range of colours	medium
polystyrene	lightweight, does not keep its shape, good heat insulator	white	low
acrylic	stiff, weather resistant, good electrical insulator	wide range of colours	high

Use the information in the table to answer the questions below.

1 Which two plastics could be best used for covering electrical cables?
2 A manufacturer is going to produce cheap green buckets to sell at large DIY stores. Which plastic should it choose? Explain your answer.
3 Give one reason why polystyrene is not used to make garden chairs. Explain your answer.

Using materials to fight crime

Forensic evidence

Forensic science is used to help fight crime. There are lots of ways in which science can help solve crimes. More and more of the evidence that is used in court to identify a criminal is scientific.

A wide variety of physical evidence can be collected at a crime scene and this can be very valuable. This includes:

▶ biological evidence (e.g. blood, bodily fluids, hair and other tissues)
▶ fingerprint evidence (or palm prints, foot prints, etc.)
▶ footwear and tyre track evidence
▶ trace evidence (e.g. clothing fibres, soil, vegetation, glass fragments)
▶ digital evidence (e.g. phone records, internet logs and email messages)
▶ drug or explosives evidence.

Trace evidence

At a crime scene, there are often tiny pieces or fragments of physical evidence such as hairs, fibres from clothing or carpets, or pieces of glass that can help work out a sequence of events. These are referred to as trace evidence, and can be transferred when two objects touch or when small particles are dispersed by an action or movement.

For example, paint can be transferred from one car to another in a collision, or a hair can be left on a sweater in a physical assault. This evidence can be used to reconstruct an event or indicate that a person or object was present.

Scientists examine the physical, optical and chemical properties of trace evidence and use a variety of instruments to find and compare samples. Most test methods require magnification and/or chemical analysis.

A surprising amount can be learned about what happened at a crime scene through trace evidence, such as whether an item or body was moved or whether someone was assaulted from behind or the side. Trace evidence can include a wide variety of materials, but the most commonly tested are hair, fibres, paint and glass. Other less frequently included items are soil, vegetation and cosmetics.

Hair and clothing fibres

There are many different types of fibres used in clothing nowadays. Someone who commits a crime could leave some clothing fibres at the scene of the crime. These can be collected and examined using a powerful microscope called a **scanning electron microscope** (Figure 12.7). They can be compared with fibres from the clothing of the suspects.

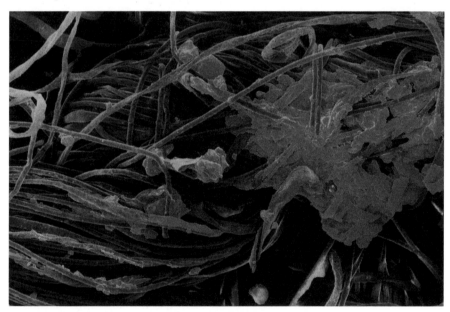

▲ **Figure 12.7** Clothing fibres viewed by means of a scanning electron microscope

Forensic scientists will look at:

▶ the mixture of fibres, for example nylon mixed with cotton
▶ the colours of the fibres, for example red fibres mixed with white
▶ the twist pattern of the fibres, for example how tightly the fibres are twisted together
▶ the weave pattern of the fibres, for example how the material is woven together
▶ stains within the fibres, for example blood, saliva or sweat.

A light microscope on high magnification can see some of these features of fibres. Hair is also a natural fibre that can give a lot of information about the criminal. Hair found at the scene can be compared with hair from a suspect in a similar way to the way that fibres are compared.

Paint and glass fragments

At a crime scene, for example, a car collision, fragments of paint and glass may be collected. These would be magnified and compared with a database of manufacturers' paint and glass. For example, every car company will use a slightly different shade or composition of red paint for their red cars, so by comparing a trace sample of paint, the make (and possibly the model) of a car involved in a car collision could be identified.

There are also many different types of glass: glass from a windscreen is a different colour and composition to a drinking glass or a window from a front door. This means that glass fragments on a suspect's clothing could be compared with those collected at a crime scene to see if that individual was present.

Fingerprints

Everyone has different patterns of lines in the skin on their fingers. These are unique, and can be used to identify an individual: the patterns leave a mark on any surface they touch. There are four main types of fingerprint ridge pattern (Figure 12.8): **arch**, **loop**, **whorl** and **composite**.

▲ **Figure 12.8** The four fingerprint pattern types

Criminals can be identified by matching their fingerprint patterns with fingerprints taken at the crime scene. This system was adopted by Scotland Yard in 1901.

Using powders to reveal fingerprints

Fingerprints are almost invisible, and forensic scientists have to dust powder over them to make them visible (Figure 12.9). The powder they use depends on the surface that the fingerprint is on (Table 12.3). Powders are used, as the solid particles are very small and can stick to the pattern, displaying it clearly.

Table 12.3 Revealing fingerprints

Surface	Powder
white	carbon black powder
black or mirrored	aluminium dust or powder

▲ **Figure 12.9** Dusting for fingerprints using carbon black powder on a white surface

Alternative light sources to reveal fingerprints

Fingerprints are not usually visible to the naked eye. It is common for scientists at a crime scene to use an alternative light source to visualise any fingerprints (Figure 12.10) and then to dust them with a suitable powder or simply photograph them. The light source may have various different coloured filters to provide a range of options.

▲ **Figure 12.10** Collecting footprints using an ultraviolet light to visualise them

Chemical developers to reveal fingerprints

Fluorescent dyes that glow under ultraviolet light can also be used as chemical developers to show up a fingerprint more clearly (Figure 12.11). If fingerprints are on a surface where they cannot be shown clearly by dusting, forensic scientists have to treat them with substances such as iodine fumes or superglue vapour to make them clearer. Iodine fumes are used to show up fingerprints on paper and clothing, and even on skin. They stain the fingerprint brown.

Collecting fingerprints for analysis

Once a fingerprint on a surface has been dusted, it has to be transferred. The excess dust is brushed off, and sticky tape is pressed over the fingerprint. The fingerprint will transfer to the tape, and the tape can then be put onto a piece of card.

White card is used for carbon black fingerprints and black card for aluminium fingerprints. Some fingerprints are photographed digitally at the scene of the crime. Others, such as those on paper or clothing, can be taken away to a laboratory.

The fingerprints are scanned into a computer that will compare them with a fingerprint database of known criminals and also with any suspects. A 16-point system is used to compare the lines and patterns on fingerprints, to make sure that the match is correct. If the match is not exact, experts will examine the fingerprints by eye to try to verify the match.

Fingerprints are **unique** to each individual person, and good fingerprint evidence means that the person can be placed at the scene of the crime. It does not mean that they actually committed the crime, so fingerprint evidence is used with other evidence to obtain a conviction. A national fingerprint **database** would hold everyone's fingerprint data. It would mean that more crimes would be solved. They would also be solved more quickly, so making better use of police time. However, some people object to a national database as they feel it is an invasion of their privacy.

▲ **Figure 12.11** A handprint appears under ultraviolet light after treatment with a chemical developer that is fluorescent

Using fingerprints to increase security

As technology advances, it is important to keep all of our digital information secure – just as we would our physical possessions. Fingerprint recognition is a simple way to ensure that only authorised individuals can access information stored on a device or in a particular room or building.

▲ **Figure 12.12** A fingerprint scanner being used to access a secure room

▲ **Figure 12.13** A fingerprint scanner being used to access a smart watch

Practice questions

1. Materials can be described as natural or synthetic. Copy Table 12.4 and place the following materials into the correct column.
   (3 marks)

   Wool   Plastic   Wood   Glass   Petrol   Cotton

Table 12.4

Natural	Synthetic

2. Most modern window frames are made from plastic rather than wood. Suggest **two** reasons why plastic is better than wood. *(2 marks)*

3. Table 12.5 below gives information about some materials.

Table 12.5

Material	Relative heaviness	Relative strength	Relative stiffness	Relative cost
steel	7800	10	105	low
kevlar	1500	30	95	very high
carbon-fibre-reinforced plastic	1600	18	100	high
glass-fibre-reinforced plastic	1900	15	10	high

Use the information in the table and your knowledge of materials to answer the following questions.

a) Describe **two** reasons why carbon-fibre-reinforced plastic has replaced traditional metals such as steel in the manufacture of golf clubs. *(2 marks)*

b) Kevlar can be used for supporting cables in the structure of suspension bridges. Describe **one** advantage and **one** disadvantage of using Kevlar instead of carbon-fibre-reinforced plastic. *(2 marks)*

c) One use for glass-fibre-reinforced plastic is to make the bodies of cars and boats. Explain why this is a suitable material. *(2 marks)*

4. Scientists are investigating more ways of using smart materials, such as photochromic and thermochromic materials, in everyday life.

a) What is a smart material? *(2 marks)*

b) What is meant by the term 'photochromic'? *(2 marks)*

c) State and explain **one** use of a photochromic material. *(2 marks)*

5. Forensic scientists collect a wide range of different types of evidence, such as fingerprint evidence, at a crime scene.

a) State **three** other types of evidence that a forensic scientist may collect at a crime scene. *(3 marks)*

b) Composite and loop are **two** types of fingerprint. Name the other two types. *(2 marks)*

c) Describe how a forensic scientist would collect and store fingerprint evidence from a black surface using powder. *(4 marks)*

d) Other than using powder, name **one** other method of visualising a fingerprint. *(1 mark)*

e) State **one** use of fingerprint recognition technology. *(1 mark)*

13 Metals and the reactivity series

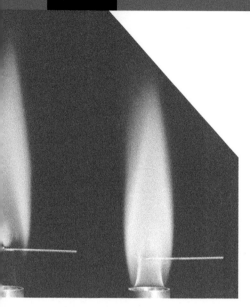

Specification points

This chapter covers sections 2.8.1 to 2.8.11 and 2.7.4 to 2.7.5 of the specification. It is about the reactions of different metals and their order of reactivity, as well as observations during reactions such as temperature change. Also covered is aluminium extraction and flame tests to identify metal ions. The required practical activities to investigate the reactivity of metals and the temperature change during a reaction are included.

The reactivity series

Over three-quarters of the elements are metals. They are found on the left-hand side of the Periodic Table.

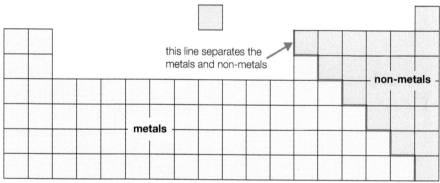

▲ **Figure 13.1** An outline of the Periodic Table showing metals on the left-hand side and non-metals on the right-hand side

The most reactive metals, such as potassium and sodium, are found in Group 1. All the other metals in the Periodic Table are less reactive than the Group 1 elements. Scientists have recorded observations of the reactions of different metals with oxygen, water and acids, and have used these to put the metals into order of their reactivity. This is called the reactivity series of metals. Part of the reactivity series is shown in Figure 13.2.

Tip

It is important to learn the reactivity series for the eight metals in Figure 13.2. You may be asked to recall it or use information from reaction observations to place them in the correct order. Make up a mnemonic to help you remember the correct order, such as Poor Scientists Can Make A Zoo In Class (the first letter of each word is the first letter of each metal in order).

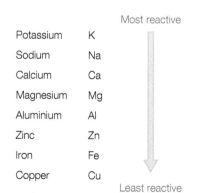

Most reactive

Potassium	K
Sodium	Na
Calcium	Ca
Magnesium	Mg
Aluminium	Al
Zinc	Zn
Iron	Fe
Copper	Cu

Least reactive

▲ **Figure 13.2** Part of the reactivity series of metals

Reaction with water

Most metals do not react with cold water. Only the most reactive metals (found at the top of the reactivity series: potassium, sodium and calcium) will react with cold water, to form a metal hydroxide and hydrogen.

metal + water → metal hydroxide + hydrogen

Example

Potassium reacts violently with water and produces a lilac flame.

potassium + water → potassium hydroxide + hydrogen

Less-reactive metals may react with hot water or steam, to produce a metal oxide and hydrogen.

metal + steam → metal oxide + hydrogen

Example

Magnesium reacts with steam.

magnesium + steam → magnesium oxide + hydrogen

Reaction with acid

Magnesium, aluminium, zinc, iron and copper are all metals. They do not react with cold water but most react with dilute acids such as hydrochloric acid. When metals react with acids, hydrogen gas is produced, so bubbles are observed.

Figure 13.3 shows three of the elements reacting with hydrochloric acid:

▶ Copper does not react with hydrochloric acid, so it is the least reactive of the three.

▶ Magnesium produces the most bubbles, so it is the most reactive.

▶ Zinc is more reactive than copper but less reactive than magnesium.

▶ If iron was included, it would be less reactive than zinc and more reactive than copper. You would see a few bubbles in the test tube.

▶ Aluminium is more reactive than zinc, iron and copper but less reactive than magnesium.

a)  b)  c)

▲ **Figure 13.3** Metals reacting with hydrochloric acid: **a)** copper, **b)** magnesium and **c)** zinc

Example

When a metal reacts with an acid, a salt and hydrogen are produced:

metal + acid → salt + water

zinc + hydrochloric acid → zinc chloride + hydrogen

magnesium + sulfuric acid → magnesium sulfate + hydrogen

See Chapter 9 to revise the names of salts.

The balanced symbol equations for these reactions are:

$$Zn + 2HCl \rightarrow ZnCl_2 + H_2$$

$$Mg + H_2SO_4 \rightarrow MgSO_4 + H_2$$

Prescribed practical

Prescribed practical C2: Investigating the reactivity of metals

Note: you will need to perform a preliminary test to find a suitable volume and concentration of acid and mass of metal to use to ensure that a sufficient volume of gas is produced.

Procedure

1 You will be given three samples of powdered metals and some hydrochloric acid.

2 Set up a water trough with a 25 cm³ measuring cylinder upturned to collect the gas produced over water, or use a gas syringe.

3 Measure the acid and place in a conical flask.

4 Weigh the first metal powder.

5 Add the metal powder to the acid, place the bung with the delivery tube into the conical flask, and start the stopwatch.

6 Record the time it takes for the metal to produce about 15 cm³ of gas (see Table 13.1).

7 Repeat the test, taking an average of the two times.

8 Repeat using a different metal.

Table 13.1 Record the time taken to produce 15 cm³ of gas

Metal	Time 1/s	Time 2/s	Average time/s

9 Draw a bar chart of time against metal using axes similar to the outline in Figure 13.4. When drawing this graph you should use a ruler to ensure neat bars; there should also be a small space between each bar.

10 Place the three metals in their order of reactivity.

Controlled variables (factors that should be kept the same) in this investigation include:

- the volume of the acid
- the type of acid
- the mass of the metal
- the size of grain of the powdered metal.

These reactions also produce heat energy, which can affect how fast the reaction is.

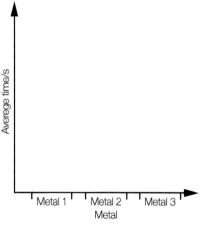

▲ **Figure 13.4**

Most of the errors in this investigation will be human errors, such as:

- not being able to start the stopwatch at the exact moment that the metal is added to the acid
- not being able to accurately judge when the required volume of gas has been produced.

To make the experiment as accurate as possible, measurements should be made using accurate apparatus such as a pipette or burette to measure the acid, and a digital scale that reads to two decimal places to weigh the metal powder.

Repeating the experiment will make the results more reliable.

Tip

You can also complete a similar investigation by measuring the loss in mass during the reaction or the time taken for the metal to disappear. This could also be a qualitative investigation where observations are recorded as each metal reacts with the acid.

Questions and sample data

1 State two ways (controlled variables) in which you made sure that this was a fair test.

2 Explain where errors may have occurred in your investigation.

3 A student carried out a similar investigation. Her results are shown in Table 13.2.

Table 13.2

Metal	Time 1/s	Time 2/s	Average time/s
zinc	16	16	16.0
iron	30	31	30.5
magnesium	10	13	11.5

a) Place the three metals (zinc, iron and magnesium) in their order of reactivity.

b) What test could the student do to identify the gas that was produced?

c) Suggest why the student did not use potassium in her investigation.

d) Suggest how the student could ensure that her results were accurate.

e) If she repeated the reaction with a stronger acid, what prediction could you make about the times taken to produce the gas?

Flame tests

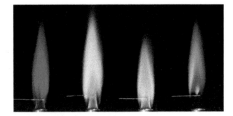

▲ **Figure 13.5** Flame tests for, from left to right, calcium, copper, potassium and sodium

Some metal ions produce very characteristic coloured flames when they are heated in a blue Bunsen flame (Table 13.3). This can be used to show the presence of these ions, and is called a flame test (Figure 13.5). A flame-test rod, usually a nichrome wire, is dipped into concentrated acid and placed in a blue Bunsen flame to clean it. It can then be put into the acid again and then into the sample and heated in the blue flame. The flame-test rod should be cleaned between each sample. Safety glasses should be worn and care taken when using a Bunsen burner.

Table 13.3 Metal ions and the colours of their flame

Ion	Flame colour
lithium (Li^+)	crimson
sodium (Na^+)	orange/yellow
potassium (K^+)	lilac
calcium (Ca^{2+})	red
copper (Cu^{2+})	blue-green

Show you can

Use the information in Table 13.4 to place the metals (X, Y and Z) in their order of reactivity, with the most reactive first.

Table 13.4

Metal	Observations on reaction of the metal with acid
X	bubbles formed slowly
Y	a large amount of bubbles produced quickly
Z	no bubbles formed

Compounds (usually chlorides) containing these ions also give these characteristic colours when heated in a flame. For example, sodium chloride will give an orange-yellow colour when heated in a roaring Bunsen flame, while copper nitrate will give a blue-green flame test colour. A flame test can be used to identify metal ions found at the scene of a crime. They can then be compared with those found on a suspect's clothing.

Test yourself

1 What does the reactivity series of metals tell us?
2 Write a word equation for the reaction of sodium with water.
3 Write a word equation for the reaction of aluminium with sulfuric acid.
4 Describe how to carry out a flame test to identify sodium and copper ions, and state the flame colours you would expect for these ions.

Energetics

Most of the reactions of metals produce heat. Any reaction that gives out heat and increases in temperature is described as exothermic. Most reactions are exothermic. Other reactions may take in heat and decrease in temperature, and these are described as endothermic.

Simple energy level diagrams (or reaction profiles) can be drawn to represent exothermic and endothermic reactions (Figure 13.6).

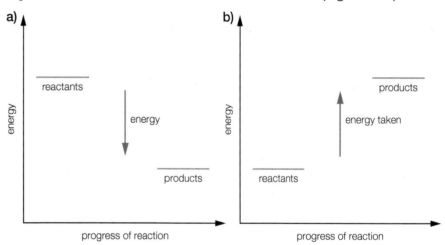

▲ **Figure 13.6** Energy level diagrams for **a)** exothermic and **b)** endothermic reactions

Prescribed practical

Prescribed practical C3: Investigating the temperature change during a reaction

Procedure

1 You will be given three samples of powered metals and some 1 M hydrochloric acid.
2 Measure 10 cm³ of acid and place this in a boiling tube.
3 Record the temperature of the acid, using a table like the one shown below (Table 13.5).
4 Weigh out 0.5 g of the first metal powder.
5 Add the metal powder to the acid.
6 Record the highest temperature reached.
7 Repeat the test, taking an average of the two temperatures.
8 Repeat using the different metals.

Table 13.5 Recording temperature changes

	Metal		
	Magnesium	Zinc	Iron
initial temperature/°C			
final temperature/°C			
temperature change/°C			
repeat initial temperature/°C			
repeat final temperature/°C			
repeat temperature change/°C			
average temperature change/°C			

The metal with the greatest temperature change is the most reactive. Place the three metals in their order of reactivity.

Controlled variables (factors that should be kept the same) in this investigation include:

- the volume of the acid
- the type of acid
- the mass of the metal
- the size of grain of the powdered metal.

To make the experiment as accurate as possible, measurements should be made with accurate apparatus such as a pipette or burette to measure the acid, a digital scale that reads to two decimal places to weigh the metal powder and a thermometer that reads to at least one decimal place (or a temperature probe).

Repeating the experiment will make the results more reliable.

Questions and sample data

1 How did you make sure that you had accurate results?
2 Why did you repeat each experiment?
3 A student carried out a similar investigation. His results are shown in Table 13.6.

Tip

You can also complete a similar investigation by using the same type of metal but different types or strengths of acids. Energy changes could also be investigated for a neutralisation reaction (as in CCEA Prescribed practical C1) by measuring the temperature rather than the pH as acid is gradually added to an alkali.

Table 13.6 Sample results

	Metal		
	1	2	3
initial temperature/°C	21	20	22
final temperature/°C	60	25	41
temperature change/°C	39	5	19
repeat initial temperature/°C	20	21	22
repeat final temperature/°C	62	26	42
repeat temperature change/°C	38	5	20
average temperature change/°C	38.5	5.0	19.5

a) Place the three metals (1, 2 and 3) in their order of reactivity.

b) The metals that the student actually used were magnesium, aluminium and iron. Suggest the identity of metal 1.

c) What would the student need to have done to ensure he had carried out a fair test?

d) If he repeated the reaction using only magnesium metal with a series of different strengths of acid, what prediction could you make about the temperatures recorded?

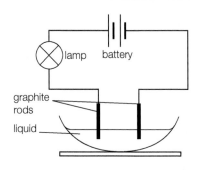

▲ **Figure 13.7** The apparatus for simple electrolysis

Electrolysis

Two graphite rods, placed in a liquid and connected externally to a power supply such as a battery or a power pack, can be used to test whether the liquid conducts electricity, as shown in Figure 13.7. If the liquid conducts electricity and is broken down or decomposed by it, then electrolysis is taking place.

The electrolyte is the liquid or solution that conducts electricity and is decomposed by it. The graphite rods used in electrolysis are called electrodes. Graphite is used because it conducts electricity and is unreactive.

The negative electrode is called the cathode. The positive electrode is called the anode.

How electrolysis works

▶ All electrolytes conduct electricity as they have free ions that can move and carry charge.

▶ When these positive and negative ions are free to move, the positive ions (called cations) move to the negative electrode (the cathode). The negative ions (called anions) move to the positive electrode (the anode). This is shown in Figure 13.8.

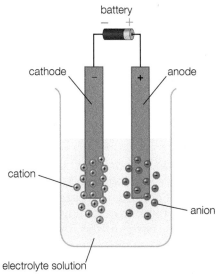

▲ **Figure 13.8** Cations and anions in solution

Extraction of aluminium from its ore

▶ Aluminium metal is extracted from its ore using electrolysis (Figure 13.9). The ore is called bauxite.

▶ Bauxite is purified to form aluminium oxide (called alumina). The alumina is dissolved in molten cryolite to reduce its melting point and increase the conductivity.

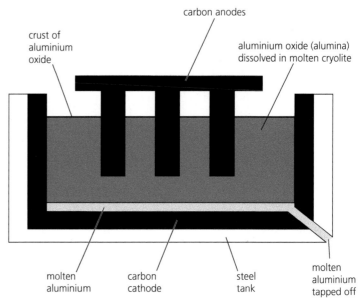

▲ **Figure 13.9** The apparatus used to extract aluminium from its ore

▶ The crust of aluminium oxide keeps heat in. The operating temperature is 900 °C. The cathode and anode are made of carbon.

▶ The positive aluminium ions move towards the negative electrode, where they gain three electrons to become aluminium atoms.

The reaction at the cathode is:

$$Al^{3+} + 3e^- \rightarrow Al$$

▶ The negative oxide ions move towards the positive electrode, where they lose two electrons to become oxygen atoms. The oxygen atoms join together to make oxygen molecules.

The reaction at the anode is:

$$2O^{2-} - 4e^- \rightarrow O_2$$

▶ The carbon anode has to be replaced periodically, as it wears away because of its reaction with oxygen. The equation for this reaction is:

$$C + O_2 \rightarrow CO_2$$

▶ The extraction of aluminium is expensive because the cost of electricity is high and a high temperature is needed to keep the aluminium oxide molten. The use of cryolite increases the conductivity and reduces the operating temperature, saving money. The aluminium oxide crust keeps some of the heat in, again saving money.

The energy required to recycle aluminium is only a fraction of the cost of producing new aluminium from bauxite. This is why it is important to recycle materials such as aluminium. It saves resources, saves energy, prevents waste going to landfill and costs less.

Practice questions

1 Copy and complete the following sentences. Metals are found on the _____ of the Periodic Table. The most reactive metals are found in Group _____. The reactivity series places metals in order of their reactivity, from most to least. The order of eight of these metals is potassium, _____, calcium, magnesium, _____, zinc, _____ and copper. *(5 marks)*

2 Reactions of metals with acid are usually exothermic.
   a) Define the term 'exothermic'. *(1 mark)*
   b) i) Describe how you would carry out an investigation to compare how exothermic the reaction of magnesium and acid is with how exothermic the reaction of zinc and acid is. Include the measurements you would take. *(4 marks)*
   ii) How would you make sure that this investigation was a fair test? *(1 mark)*
   iii) How would you make sure that the results were accurate? *(1 mark)*
   iv) How would you make sure that the results were reliable? *(1 mark)*
   v) What result would you expect? *(1 mark)*

3 A student carried out an investigation into the energy changes in a neutralisation reaction. She added sodium hydroxide to $10\,cm^3$ of hydrochloric acid, $1\,cm^3$ at a time, and recorded the temperature each time. The acid and alkali were both the same concentration. The results are shown in Table 13.7.

Table 13.7

Volume of sodium hydroxide added/cm^3	Temperature/°C
0	22.0
1	23.0
2	24.5
3	25.0
4	26.0
5	27.5
6	28.0
7	29.0
8	30.5
9	31.5
10	33.0
11	31.0
12	28.5

   a) State one safety precaution that the student should have taken. *(1 mark)*
   b) Name a piece of apparatus that the student should have used to accurately measure the acid. *(1 mark)*
   c) i) Describe the trend in the results from $0\,cm^3$ to $10\,cm^3$. *(1 mark)*
   ii) Describe the trend in the results from $10\,cm^3$ to $12\,cm^3$. *(1 mark)*
   iii) Explain the trends in i) and ii). *(2 marks)*
   d) The student carried out the experiment in a polystyrene cup surrounded by cotton wool. Suggest a reason for this set-up. *(1 mark)*
   e) Suggest what would happen to the results if the student used more concentrated acid. *(1 mark)*

4 Flame tests can be used to identify the metal ion in a compound.
   a) Copy and complete Table 13.8 about flame colours. *(3 marks)*

Table 13.8

Metal ion	Flame colour
lithium	crimson
	orange/yellow
potassium	
	blue-green

   b) The steps required to carry out a flame test are shown below, but they are not in the correct order.
   1 Dip the flame-test rod (nichrome wire) into concentrated acid.
   2 Heat the sample in the blue flame and record the colour.
   3 Place the rod in a blue Bunsen flame to clean it.
   4 Place the rod into the acid again and then into the sample.
   Place the steps (1, 2, 3 and 4) in the correct order (the first one has been done for you).
   1 __ __ __ *(2 marks)*

5 Electrolysis can be used to extract aluminium from its ore.

a) Define the term 'electrolysis'. *(2 marks)*

b) Figure 13.10 shows the set-up for the electrolysis of aluminium ore.

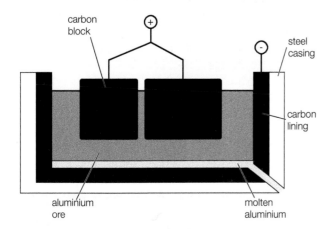

Figure 13.10

i) Copy and complete the following sentences about the extraction of aluminium.
The ore of aluminium is called _____.
This is purified before electrolysis, to form aluminium oxide. The aluminium oxide is dissolved in molten _____ to reduce the melting point. The carbon lining acts as the negative electrode, which is called the _____. The carbon blocks are the positive electrode, called the _____. *(4 marks)*

ii) Write a balanced symbol equation for the formation of aluminium at the negative electrode. *(3 marks)*

iii) Suggest one reason why the carbon blocks need to be replaced periodically. *(1 mark)*

c) Give two reasons why it is important to recycle aluminium. *(2 marks)*

14 Rates of reaction

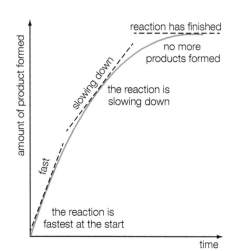

Specification points

This chapter covers sections 2.9.1 to 2.9.7 of the specification. It is about rates of reaction, how to change the speed of a reaction and how to measure the rate of a reaction.

Rates of reaction

Some reactions happen quickly, for example an explosion, while some happen more slowly, for example a nail rusting. The speed at which a reaction happens is called the rate of the reaction. It can be measured in terms of the change in the amount of reactants or products with time.

Tip

When describing a rate of reaction it is always important to include **time**, for example, the rate of a reaction could be described as how quickly the reactants are used up **per unit time**, or **per second**.

It is relatively straightforward to measure the rate of a simple reaction: you can either measure how quickly the products are formed or how quickly the reactants are used up. For a simple reaction, such as magnesium metal reacting with hydrochloric acid, the rate of the reaction will change during the reaction:

▶ At the beginning of the reaction the rate will be relatively **fast**.

▶ It will then **slow down** as the reactants are being used up.

▶ Finally it will **stop** when all the reactants are used up.

This means that we can either describe the rate of reaction at a specific time or describe the average or mean rate of the entire reaction.

To calculate the rate of a reaction, the amount of reactant used up or product formed is divided by the time taken.

$$\text{Rate} = \frac{\text{reactant used or product formed}}{\text{time}}$$

If the time is measured in seconds (s) then the unit for the rate is s^{-1}.

Graphs can be drawn to show the change in the amount of reactant or product with time (see Figure 14.1). The shapes of these graphs are usually similar. The slope of the line represents the rate of the reaction. The steeper the slope, the faster the reaction.

▲ **Figure 14.1** Graph showing the progress of a reaction

For a simple reaction such as magnesium metal reacting with hydrochloric acid the slope of the graph will change during the reaction:

▶ At the beginning of the graph the slope will be **steepest**, when the reaction is fastest.

▶ It will become **less steep** as the reaction slows down.

▶ Finally the line will become **horizontal** as the reaction stops.

It is easiest to follow the rate of a reaction when the reaction produces a gas: for example, a metal and acid will produce hydrogen gas, or a metal carbonate with an acid will produce carbon dioxide (Figure 14.2).

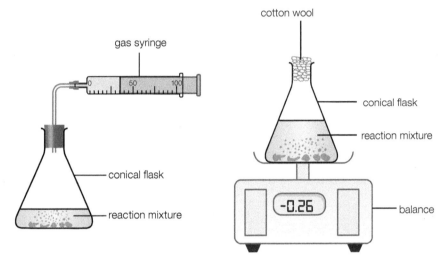

▲ **Figure 14.2** Two ways of measuring the amount of gas produced during a reaction

Factors affecting the rate of a reaction

There are several factors that can affect the rate of a chemical reaction. To explain the effect of these factors, we should first understand how a reaction happens.

For a chemical reaction to happen, the reactant particles must collide with each other with enough energy to cause them to react as shown in Figure 14.3. The minimum amount of energy required for a successful collision to occur is called the activation energy. This explanation of how a reaction happens is called collision theory.

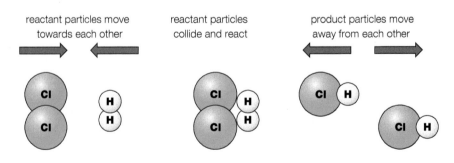

▲ **Figure 14.3** A successful collision

Temperature

The higher the temperature of a reaction, the faster the rate of reaction. A small increase in temperature can have a big effect on the rate of reaction.

As the temperature of a reaction increases, the reactant particles have more energy, and so more of the particles will have enough energy to react when they collide. The particles are also moving faster at higher temperatures, and so will collide more often. Both of these things lead to more frequent successful collisions so there is a faster rate of reaction.

Concentration

The higher the concentration of the reactants in a solution, the faster the rate of reaction. The rate of a reaction is proportional to the concentration of the reactants.

When the concentration of a reactant increases, the number of particles in the reactant solution also increases. If there are more particles, then they are likely to collide more often. This leads to more frequent successful collisions and so there is a faster rate of reaction.

Using a catalyst

A catalyst is a substance that changes (usually increases) the rate of a reaction without being used up. Catalysts work by providing a different pathway for a reaction that has a lower activation energy. This means that more reactant particles will have enough energy for a successful collision.

Different reactions have different catalysts. Manganese dioxide catalyses the breakdown of hydrogen peroxide to produce oxygen. Enzymes are an example of a catalyst in biology: for example, amylase is the enzyme that catalyses the breakdown of starch into sugars.

Since a catalyst is not used up during a reaction, the same amount is present at the start and end of a reaction. This means that there is no need for it to appear in the chemical equation. Sometimes the catalyst appears above the arrow in the equation:

$$\text{hydrogen peroxide} \xrightarrow{\text{manganese dioxide}} \text{water} + \text{oxygen}$$

The effect of changing conditions on the rate of a reaction can be investigated in many different ways. It is most easily done a gas is produced during the reaction. The volume or mass of gas can be measured over time, and a graph can be drawn to show the results.

Test yourself

1 Define the term 'rate of reaction'.
2 State two reaction conditions that can change the rate of reaction.
3 A student measured the volume of gas produced at regular intervals when a metal reacted with hydrochloric acid.
   a) Name the gas produced.
   b) i) Sketch the shape of the graph that the student should expect, showing the volume of gas formed against time for the reaction.
      ii) Explain what the shape of the graph shows.

Show you can

Carbon dioxide is formed when copper carbonate reacts with sulfuric acid. Table 14.1 shows how the volume of carbon dioxide changed with time when some copper carbonate was reacted with sulfuric acid.

Table 14.1

Time/s	0	10	20	30	40	50	60	70	80
Volume of carbon dioxide/cm³	0	15	23	37	44	49	51	52	52

Plot a graph of the volume of carbon dioxide against time.

Calculate the average rate of the overall reaction after 70 seconds by using the equation:

$$\text{rate} = \frac{\text{volume of gas}}{\text{time taken}}$$

Practical activity

Investigating the progress of a reaction

Method 1: Using a gas syringe

For example, magnesium reacting with hydrochloric acid produces hydrogen gas.

To measure the volume of gas produced during a reaction, a gas syringe can be attached to the reaction vessel (e.g. a conical flask), as shown in Figure 14.4. The volume of gas is recorded at regular intervals. Alternatively, the gas syringe plunger can be attached to a movement sensor and a computer, and the readings taken electronically – this would give more accurate results.

A graph of the volume of gas against time can be plotted to illustrate the progress of the reaction. You can find a numerical value for the rate by dividing the total volume of gas produced by the time taken for the reaction to finish.

Method 2: Using a balance

For example, calcium carbonate reacting with hydrochloric acid produces carbon dioxide gas.

To measure the mass of gas produced during a reaction, the reaction vessel (e.g. a conical flask) can be placed on a balance, as shown in Figure 14.5. The balance is set to zero, when the apparatus and chemicals are on the balance, before the reaction starts. The mass of the reaction is recorded at regular intervals. The cotton wool is placed in the neck of the conical flask to prevent any chemicals splashing out. The gas produced can escape through the cotton wool, and the mass of the reaction decreases as it does so.

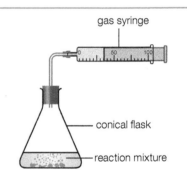

▲ **Figure 14.4** Measuring the volume of gas formed using a gas syringe

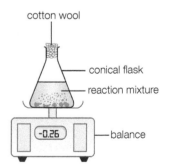

▲ **Figure 14.5** Measuring the mass of gas formed during a reaction using a balance

A graph of the mass of gas against time can be plotted to illustrate the progress of the reaction. A numerical value for the rate can be obtained by dividing the total mass of gas produced by the time taken for the reaction to finish.

Method 3: Measuring a solid forming

For example, sodium thiosulfate reacting with hydrochloric acid produces the solid sulfur.

If a reaction produces an insoluble solid, then as the reaction takes place the solution will become cloudy. To measure the rate of such a reaction, the reaction vessel (e.g. a conical flask) can be placed on a piece of paper with a cross drawn on it, as shown in Figure 14.6. As the solid forms, the solution becomes cloudy, and the time taken for the cross to no longer be visible is measured. Alternatively, a light sensor and a computer could be used, and the readings taken electronically – this would give more accurate results. This method will only measure the time taken for the reaction to be complete – it cannot be used to follow the progress of a reaction as in methods 1 and 2.

By finding the reciprocal of the time taken for the reaction to finish (1/time), a numerical value for the rate can be obtained.

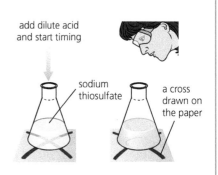

▲ **Figure 14.6** Measuring an insoluble solid forming during a reaction

Using rate graphs to compare reactions

If several experiments are carried out to find the rate of the same reaction under different conditions, then rate graphs can be drawn to compare the results. For example, if zinc was added to sulfuric acid and the volume of gas formed was measured at regular intervals, a graph could be plotted; this may look like the one in Figure 14.7. If the reaction was then repeated at increasing concentrations of acid and the lines were plotted on the same axes, then the graph may look like the one in Figure 14.8.

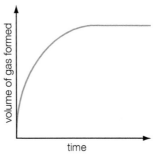

▲ **Figure 14.7** Volume of gas formed during the reaction

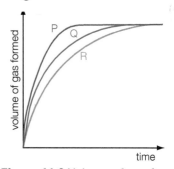

▲ **Figure 14.8** Volume of gas formed during the reaction at three different concentrations P, Q and R

In Figure 14.8, the lines labelled Q and P represent the reactions being carried out at higher concentrations. This is seen in the graph because the lines are steeper. It is important to note that all the reactions produce the same volume of gas at the end of the reaction – this is because the same amount of reactants was used each time and so the same amount of products is produced. The only difference will be the time (or rate) at which the reaction finishes.

1 The rate of a reaction can be affected by several reaction conditions. State what happens to the rate of a reaction when the temperature increases. *(1 mark)*

2 A student carried out four experiments to follow the rate of the reaction of calcium carbonate with hydrochloric acid. He used four different concentrations of acid (0.5 mol/dm³, 1.0 mol/dm³, 1.5 mol/dm³ and 2.0 mol/dm³). He used his results to plot a graph, which is shown in Figure 14.9.

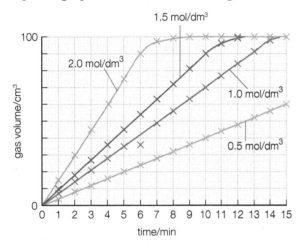

Figure 14.9

   a) The student used a gas syringe to get his measurements. Sketch and label a diagram of the apparatus set-up required for this experiment. *(4 marks)*
   b) Suggest two ways in which the student could make this a fair test. *(2 marks)*
   c) Look at the graph.
      i) Identify the one anomalous result on the graph. *(1 mark)*
      ii) Which concentration of acid produced the slowest rate of reaction? Explain why you chose this answer. *(2 marks)*
      iii) Copy and complete the following sentence about the trend shown in the graph.
         As the concentration of the acid increases,
         _____. *(1 mark)*
      iv) Explain the trend in terms of collision theory. *(2 marks)*

3 A student used the apparatus shown in Figure 14.10 to follow the reaction of hydrogen peroxide decomposing in the presence of solid manganese(IV) oxide. The reaction produces water and oxygen.

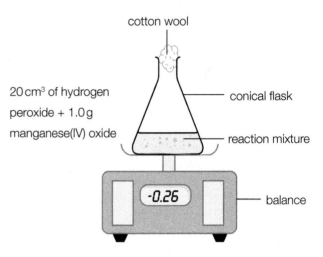

Figure 14.10

   She recorded the mass of oxygen lost every minute. Her results are shown in Table 14.2.

Table 14.2

Time/min	0	1	2	3	4	5	6	7
Mass of oxygen lost/g	0.0	0.23	0.34	0.35	0.45	0.47	0.48	0.48

   a) Look at the apparatus set-up. What is the purpose of the cotton wool? *(1 mark)*
   b) i) Plot a line graph of the mass of oxygen lost (g) against time (min). *(3 marks)*
      ii) Identify the anomalous result. *(1 mark)*
      iii) Describe and explain the shape of the graph. *(3 marks)*
   c) Suggest how the student could make sure that the results are reliable. *(1 mark)*
   d) Suggest an alternative experiment to measure the rate of this reaction. *(2 marks)*
   e) If this same experiment was carried out at a higher temperature, suggest one similarity and one difference between the results obtained. *(2 marks)*

4 a) Explain, using collision theory, how each of the following speeds up the rate of a reaction:
      i) an increase in temperature *(3 marks)*
      ii) an increase in concentration. *(3 marks)*
   b) A catalyst can also affect the rate of reaction. Define the term 'catalyst'. *(2 marks)*

Organic chemistry

Organic chemistry is the study of molecules containing carbon. There are a huge number of carbon compounds, so scientists study them by looking at different 'families' of compounds that have similar structures or properties. These families are called homologous series.

Each homologous series has:

▶ the same general formula
▶ similar chemical properties
▶ a gradual change in physical properties

Each member of a homologous series differs by $-CH_2$.

You will need to know about two different homologous series: alkanes and alkenes. Both of these homologous series are hydrocarbons. A hydrocarbon is a chemical that contains only the elements carbon and hydrogen. Many hydrocarbons come from crude oil.

Crude oil

Crude oil is a mixture of liquid hydrocarbons, with other gases and solid hydrocarbons dissolved in it. It is a finite energy resource – this means that there is only a certain amount of it, which cannot quickly be replaced, and therefore, someday, crude oil will run out. Crude oil has been formed over millions of years by the action of heat and pressure on dead plants and animals, as shown in Figure 15.1.

It is not easy to measure the amount of crude oil left in the world. It has to be estimated. The four main factors that affect this sort of estimate are discussed below.

▶ Improved methods of extraction would allow for more oil to be taken from a reserve than was previously possible.
▶ New oil reserves are being discovered, as methods of detecting them are getting better.
▶ Our use of crude oil is changing. We would use less oil if we increased the amount of plastic recycling, or if we cut down on the amount of petrol and diesel that we use.

▶ Government projections on the use of crude oil may not be accurate, as they may change the numbers for political reasons:

- If a government wants to seem more environmentally friendly, it can make the figures for the use of crude oil and projections for future use appear lower.
- If a government's opponents want to make the government seem less environmentally friendly, they can make the figures for the use of crude oil and projections for future use appear higher.

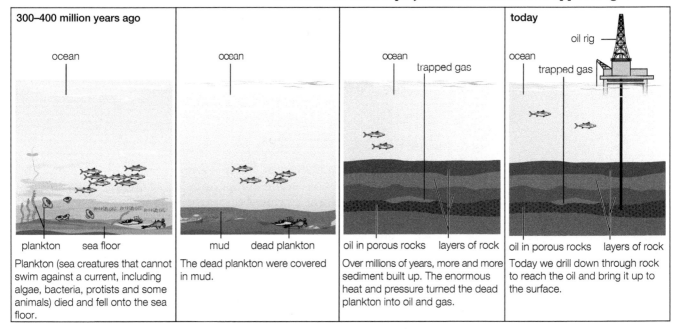

▲ **Figure 15.1** The formation and extraction of crude oil

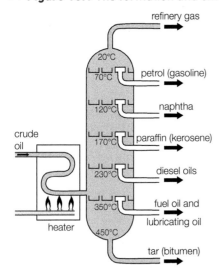

▲ **Figure 15.2** Fractional distillation of crude oil

Separating crude oil

Crude oil is a mixture. It can be separated into its different constituents by fractional distillation (Figure 15.2).

The crude oil is heated until it becomes a gas. It is then fed into a steel tower. The tower is hotter at the bottom and cooler at the top. As the mixture of gases moves up the tower, some of the gases condense, changing from a gas to a liquid. These liquids form at different levels because they have different boiling points, and are then collected separately.

The mixture of solids dissolved in crude oil is called **tar** or **bitumen**. It comes out of the bottom of the fractional distillation tower. The mixture of gases dissolved in crude oil is referred to as **refinery gases**. They do not condense. The refinery gases come out of the top of the tower, still as gases.

The simpler mixtures of hydrocarbons, mostly alkanes, produced in this way are called fractions. The refinery gases make up a fraction, and bitumen is also a fraction. Some of the other fractions are shown in Figure 15.2.

All of the fractions from crude oil contain hydrocarbons, and they are all fossil fuels. However, some would not be very good fuels. For example, bitumen will only burn at a very high temperature.

Alkanes

The simplest homologous series is the **alkanes**. They are hydrocarbons, which means that they contain only the elements hydrogen and carbon. They consist of a central chain of carbon atoms linked by only **single** covalent bonds.

Naming hydrocarbons

Alkanes and alkenes are named like many other organic compounds.

1 Count the number of carbons in the longest chain and use the correct prefix (word beginning), as shown in Table 15.1.

Table 15.1 Prefixes for the first four hydrocarbons

Number of carbon atoms	Prefix
1	meth-
2	eth-
3	prop-
4	but-

2 Then add the ending -ane for an alkane or -ene for an alkene.

Writing the formulae for alkanes

The general formula of an alkane is C_nH_{2n+2}. Their general formula tells you the relationship between the number of carbon and hydrogen atoms in each alkane. So, if there is only one (n) carbon atom, then there will be four hydrogen atoms ($2(1) + 2 = 4$). If there are two (n) carbon atoms, then there will be six hydrogen atoms ($2(2) + 2 = 6$) and so on.

Table 15.2 gives the names, chemical formulae and structural formulae of methane, ethane, propane and butane, which are the first four members of the alkanes.

Table 15.2 The first four alkanes

Name	Molecular formula	Structural formula	State at room temperature and pressure
methane	CH_4	H—C—H with H above and below	gas
ethane	C_2H_6	H—C—C—H with H above and below each C	gas
propane	C_3H_8	H—C—C—C—H with H above and below each C	gas
butane	C_4H_{10}	H—C—C—C—C—H with H above and below each C	gas

Combustion of alkanes

All hydrocarbons can be burned – they undergo combustion. When hydrocarbon fuels burn in a plentiful supply of oxygen, the hydrogen reacts with oxygen to form water, and the carbon reacts with oxygen to form carbon dioxide.

hydrocarbon + oxygen → carbon dioxide + water

This is described as complete combustion.

(In a limited supply of oxygen, carbon monoxide is produced rather than carbon dioxide, and this is called incomplete combustion.)

Natural gas is mainly made up of the hydrocarbon methane (CH_4). It is used to heat homes and for cooking. Natural gas also contains a few other hydrocarbons in small amounts.

When methane burns, the word equation for the reaction is:

methane + oxygen → carbon dioxide + water

All other hydrocarbons burn in the same way, making carbon dioxide and water. The process of burning hydrocarbons releases energy as heat.

Bottled gas (Figure 15.3) is mainly butane (C_4H_{10}), which is a hydrocarbon. When butane burns, it forms carbon dioxide and water, releasing heat:

butane + oxygen → carbon dioxide + water

▲ **Figure 15.3** Bottled butane gas

Writing balanced symbol equations for the combustion reactions of alkanes

To balance a complete combustion equation:

1 Write the chemical formula of the hydrocarbon on the left and put '+ O_2' after it. Then put an arrow (→).
2 Write 'CO_2 + H_2O' to the right of the arrow.
3 The number of carbon atoms in the hydrocarbon is the same as the balancing number in front of the CO_2.
4 The number of hydrogen atoms in the hydrocarbon is divided by 2 to get the balancing number in front of the H_2O.
5 Count the total number of oxygen atoms in CO_2 and H_2O (remember that each CO_2 molecule contains two oxygen atoms). Divide this total by 2 to get the balancing number in front of O_2.
6 If the balancing number in front of O_2 has a half, for example '2½', this can be left as a fraction, or you can multiply all the balancing numbers by 2 to get whole numbers.

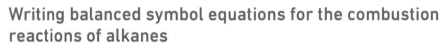

Example

Example 1
Write a balanced symbol equation for the combustion of methane (CH_4).

Step 1 $CH_4 + O_2 →$

Step 2 $CH_4 + O_2 → CO_2 + H_2O$

Step 3   There is one C atom in CH_4. This means '1' is the balancing number in front of CO_2. No number is needed, as CO_2 in an equation means $1CO_2$.

$CH_4 + O_2 → CO_2 + H_2O$

Step 4   There are four H atoms in CH_4. Dividing this by 2 gives '2' as the balancing number in front of H_2O.

$CH_4 + O_2 \rightarrow CO_2 + 2H_2O$

Step 5   There are two O atoms in CO_2 and one in each of the $2H_2O$, so there are four O atoms in total on the right-hand side of the arrow. Dividing this by 2 gives '2' to go in front of the O_2 on the left-hand side.

$CH_4 + 2O_2 \rightarrow CO_2 + 2H_2O$

Example 2

Complete the balanced symbol equation for the complete combustion of butane (C_4H_{10}).

Step 1 $C_4H_{10} + O_2 \rightarrow$

It is common for step 1 to be given in the question and for you to complete and balance the combustion equation.

Step 2 $C_4H_{10} + O_2 \rightarrow CO_2 + H_2O$

Step 3   There are four C atoms in C_4H_{10}, so 4 is the balancing number in front of CO_2.

$C_4H_{10} + O_2 \rightarrow 4CO_2 + H_2O$

Step 4   There are ten H atoms in C_4H_{10}. Dividing this by 2 gives 5 as the balancing number in front of H_2O.

$C_4H_{10} + O_2 \rightarrow 4CO_2 + 5H_2O$

Step 5   There are eight O atoms in $4CO_2$ and one in each of the $5H_2O$, so there are 13 O atoms in total on the right-hand side of the arrow. Dividing this by 2 gives 6½ to go in front of the O_2 on the left-hand side.

$C_4H_{10} + 6½O_2 \rightarrow 4CO_2 + 5H_2O$

Step 6   The above equation is perfectly correct but, normally, whole-number balancing numbers are used. Therefore, the balancing numbers are all multiplied by 2 (remember that C_4H_{10} means $1C_4H_{10}$, so it becomes $2C_4H_{10}$ when multiplied by 2).

$2C_4H_{10} + 13O_2 \rightarrow 8CO_2 + 10H_2O$

Test yourself

1   Define the term 'hydrocarbon'.
2   Copy and complete the sentences about separating crude oil.
    The process of separating crude oil is called _____ _____.
    First the crude oil is _____ until it becomes a gas. It is then fed into a steel tower. The tower is hotter at the bottom and cooler at the top. As the mixture of gases moves up the tower, some of the gases _____, changing from a gas to a liquid. These liquids form at different levels because they have different _____ _____, and are then collected separately. Each part that is collected is called a fraction.
3   One group of hydrocarbons is the alkanes.
    a) What is the general formula for an alkane?
    b) The first member of the alkane family is methane.
        i)   Give the formula of methane.
        ii) Write a word equation for the combustion of methane.
        iii) Write a symbol equation for the combustion of methane.

Show you can

Write balanced symbol equations for the complete combustion of ethane and propane.

147

Alkenes

Another homologous series containing hydrocarbons is the **alkenes**. They are hydrocarbons, which means they contain only the elements hydrogen and carbon. The simple members consist of a central chain of carbon atoms containing **one** double covalent bond.

Alkenes are named in a similar way to alkanes, by using the correct prefix (as shown in Table 15.1 on page 145) to indicate the number of carbon atoms and then adding the suffix '-ene'. For example, an alkene containing three carbon atoms will be named propene.

Writing the formulae for alkenes

The general formula of an alkene is C_nH_{2n}. Their general formula tells you the relationship between the number of carbon and hydrogen atoms in each alkene. So, if there are two (n) carbon atoms, then there will be four hydrogen atoms ($2 \times 2 = 4$). If there are four (n) carbon atoms, then there will be eight hydrogen atoms ($2 \times 4 = 8$) and so on. For GCSE Single Award Science it does not matter where the double bond is placed in the carbon chain, but it is important that each carbon atom is drawn showing only four bonds.

Table 15.3 gives the names, chemical formulae and structural formulae of ethene, propene and butene, which are the first three members of the alkenes. ('Methene' does not exist, as there cannot be a double bond if there is only one carbon atom.)

Table 15.3 The first three alkenes

Name	Molecular formula	Structural formula	State at room temperature and pressure
ethene	C_2H_4		gas
propene	C_3H_6		gas
butene	C_4H_8		gas

Combustion of alkenes

All hydrocarbons can be burned. When hydrocarbon fuels burn in a plentiful supply of oxygen, the hydrogen reacts with oxygen to form water, and the carbon reacts with oxygen to form carbon dioxide. The process of burning hydrocarbons releases energy as heat.

hydrocarbon + oxygen → carbon dioxide + water

This is described as **complete combustion**.

(In a limited supply of oxygen, carbon monoxide is produced rather than carbon dioxide and this is called incomplete combustion.)

For example, when ethene burns it forms carbon dioxide and water, releasing heat:

ethene + oxygen → carbon dioxide + water

Writing balanced symbol equations for combustion reactions of alkenes

For alkenes, you follow the same steps as for the alkanes.

Example

Example 1

Complete the balanced symbol equation for the complete combustion of ethene (C_2H_4).

Step 1 $C_2H_4 + O_2 →$

It is common for step 1 to be given in the question and for you to complete and balance the combustion equation.

Step 2 $C_2H_4 + O_2 → CO_2 + H_2O$

Step 3   There are two C atoms in C_2H_4, so 2 is the balancing number in front of CO_2.

$C_2H_4 + O_2 → 2CO_2 + H_2O$

Step 4   There are four H atoms in C_2H_4. Dividing this by 2 gives 2 as the balancing number in front of H_2O.

$C_2H_4 + O_2 → 2CO_2 + 2H_2O$

Step 5   There are four O atoms in $2CO_2$ and one in each of the $2H_2O$, so there are six O atoms in total on the right-hand side of the arrow. Dividing this by 2 gives 3 to go in front of the O_2 on the left-hand side.

$C_2H_4 + 3O_2 → 2CO_2 + 4H_2O$

Example 2

Write a balanced symbol equation for the combustion of propene (C_3H_6).

Step 1 $C_3H_6 + O_2 →$

Step 2 $C_3H_6 + O_2 → CO_2 + H_2O$

Step 3   There are three C atoms in C_3H_6. This means that 3 is the balancing number in front of CO_2.

$C_3H_6 + O_2 → 3CO_2 + H_2O$

Step 4   There are six H atoms in C_3H_6. Dividing this by 2 gives 3 as the balancing number in front of H_2O.

$C_3H_6 + O_2 → 3CO_2 + 3H_2O$

Step 5   There are two O atoms in each of the $3CO_2$, making six for the CO_2, and one in each of the $3H_2O$, so there are nine O atoms in total on the right-hand side of the arrow. Dividing this by 2 gives 4½ to go in front of the O_2 on the left-hand side.

$C_3H_6 + 4½O_2 → 3CO_2 + 3H_2O$

Step 6   The above equation is perfectly correct but, normally, whole-number balancing numbers are used. Therefore, the balancing numbers are all multiplied by 2 (remember that C_3H_6 means $1C_3H_6$, so it becomes $2C_3H_6$ when multiplied by 2).

$2C_3H_6 + 9O_2 → 6CO_2 + 6H_2O$

Atmospheric pollution

The combustion of any hydrocarbon fuel is a major source of atmospheric pollution. This is because during the complete combustion of a hydrocarbon fuel, carbon dioxide is produced. Carbon dioxide is the main greenhouse gas.

The greenhouse effect

The greenhouse effect in itself is not harmful; in fact, it is important to keep the temperature of the Earth steady.

The main greenhouse gas, carbon dioxide, builds up in the atmosphere, forming a layer like a blanket around the Earth. When the radiation (heat) from the Sun travels to the Earth, some is absorbed by the Earth and some is reflected back into the atmosphere. Some of the reflected radiation passes back into space, but some is also reflected back to the Earth by the layer of greenhouse gases. As well as carbon dioxide, other greenhouse gases include methane and water vapour. The greenhouse effect is illustrated below in Figure 15.4.

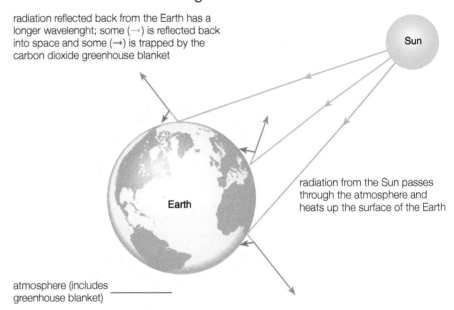

radiation reflected back from the Earth has a longer wavelenght; some (→) is reflected back into space and some (→) is trapped by the carbon dioxide greenhouse blanket

Sun

Earth

radiation from the Sun passes through the atmosphere and heats up the surface of the Earth

atmosphere (includes _____ greenhouse blanket)

▲ **Figure 15.4** The greenhouse effect

Global warming

Global warming occurs when the layer of carbon dioxide and other greenhouse gases in the Earth's atmosphere becomes thicker and so traps more of the Sun's radiation, causing a warming of the atmosphere.

Effects of global warming

The warming of the atmosphere causes:

▶ climate change – more weather extremes such as droughts and severe storms
▶ the polar ice caps to melt – and as a result sea levels will rise
▶ increased flooding
▶ increased spread of tropical diseases.

Reducing global warming

We may not be able to stop global warming – but it is important to act now to reduce global warming. Humans are responsible for producing most of the greenhouse gases, so it is up to us to accept this and try to reduce them. We could:

▶ plant more trees (these will take in carbon dioxide)
▶ reduce deforestation (by cutting down fewer trees)
▶ burn less fossil fuels (by using alternative fuels or renewable resources such as solar or wind power)
▶ be more energy efficient (by turning off lights when not in use, turning down the thermostat, etc).

> ### Test yourself
>
> 4 Alkanes and alkenes are both hydrocarbons. State one way in which an alkene is different from an alkane.
> 5 One group of hydrocarbons is the alkenes.
>   a) What is the general formula for an alkene?
>   b) The first member of the alkene family is ethene.
>   i) Give the formula of ethene.
>   ii) Write a word equation for the combustion of ethene.
>   iii) Write a symbol equation for the combustion of ethene.
> 6 Burning both alkanes and alkenes produces air pollution.
>   a) Describe the process of global warming.
>   b) State two negative effects of global warming on the environment.
>   c) Suggest two ways in which global warming could be reduced.

Polymers

A polymer is a chain of many small molecules that have been joined or bonded together (Figure 15.5). The small molecules are called monomers.

Polymerisation is the name given to the process of joining many small molecules (monomers) together to form long chains.

Plastics are an example of polymers. The monomers for many plastics come from crude oil: for example, polythene is made from ethene.

One type of polymerisation is addition polymerisation. In this process, the monomers must have a carbon–carbon double bond, and the monomers then simply 'add together'. Two examples of addition polymers are polythene and polyvinyl chloride (PVC).

> ### Tip ↺
>
> It is important to describe the process as addition polymerisation. Using the term 'additional' instead of 'addition' will not gain marks in the examination.

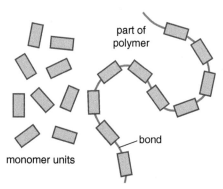

▲ **Figure 15.5** Monomers and polymers

Using ethene to produce polythene

Many ethene molecules can join together to form the common polymer polythene, which is used to make plastic bags and bottles. This is done under high temperature and pressure.

During the reaction, the C=C double bond in ethene breaks or opens, and then another molecule of ethene can add on. Typically, a polymer will have several hundred or even thousands of monomers joined together. The process can be described using the symbol equation in Figure 15.6.

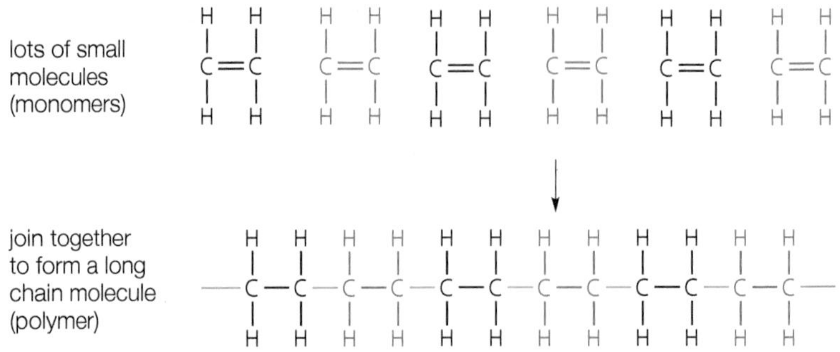

▲ **Figure 15.6** The symbol equation for making polythene

As can be seen in Figure 15.6, the polymer is made up of a unit that repeats many times – this is known as the repeating unit. (This repeating unit can be shown inside square brackets, as shown in Figure 15.7a.)

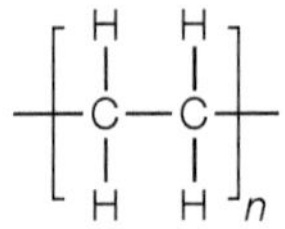

▲ **Figure 15.7** A repeating unit of polythene

Using chloroethene to produce polyvinyl chloride

Many chloroethene (old name: vinyl chloride) molecules can join together to form the common polymer polyvinyl chloride (PVC), which is used to make water pipes and the covering of window frames and doors, as well as clothing. This addition polymerisation is done under high temperature and pressure.

A chloroethene molecule is similar to an ethene molecule except that one of the hydrogen atoms has been replaced by a chlorine atom.

During the reaction, the C=C double bond in chloroethene breaks or opens, and then another molecule of chloroethene can add on. Typically, a polymer will have several hundred or even thousands of monomers joined together. The process can be described using the symbol equation in Figure 15.8.

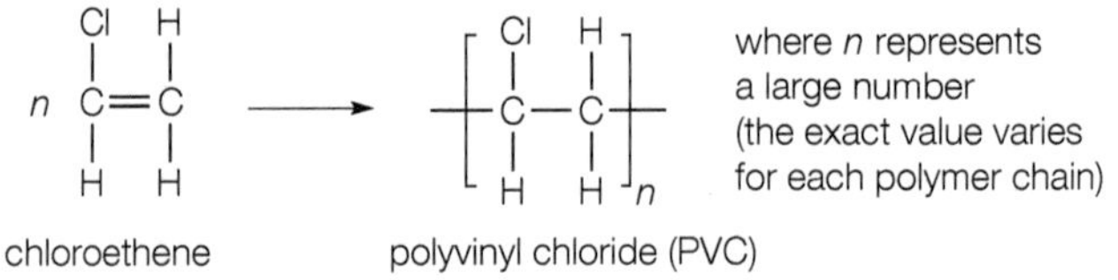

chloroethene   polyvinyl chloride (PVC)

▲ **Figure 15.8** The symbol equation for making polyvinyl chloride (PVC)

Disposing of addition polymers (plastics)

Most plastics are non-biodegradable. This means that they will not be decomposed (broken down) by microbes, unlike other waste that is biodegradable such as food and newspapers.

Most waste is disposed of by sending it to landfill, where it is buried. If the waste is biodegradable, then it will eventually rot away and take up less space.

The advantage of landfill is that it is convenient and all waste can be disposed of there. The disadvantages are that it takes up a lot of space, and some areas are running out of space for landfill. Also, as some of the waste decomposes, foul-smelling liquids and gases can be produced, some of which seep into the ground.

▲ **Figure 15.9** Landfill

Another way to dispose of waste is by incineration. This means that the waste is burned. The advantage of this is that it is relatively quick and does not take up land space, like landfill. The main disadvantage is the large amount of carbon dioxide produced when the waste is burned. Carbon dioxide contributes to global warming.

▲ **Figure 15.10** A waste incineration plant

▲ **Figure 15.11** A waste incineration kiln

Some scientists are investigating the use of biodegradable plastics as an alternative to more traditional plastics. It is also important, where possible, to recycle plastics.

1 Copy and complete Table 15.4 about the first three alkenes. *(4 marks)*

Table 15.4

Name	Molecular formula	Structural formula	State at room temperature and pressure
Ethene	C_2H_4	H∖C=C∕H ⁄ ∖ H H	
	C_3H_6		gas
butene		H H H H H−C−C−C=C ∖ H H H H	gas

2 a) i) Copy and complete the word equation for the combustion of methane.

methane + _____ → carbon dioxide + _____ *(2 marks)*

ii) Write a balanced symbol equation for the combustion of methane. *(3 marks)*

iii) One of the products of the combustion of methane is carbon dioxide. Carbon dioxide is the main greenhouse gas, which contributes to global warming. Describe how carbon dioxide causes global warming and the effects it has on the environment. *(6 marks)*

b) Another group of hydrocarbons is the alkenes.

i) What is meant by the term 'hydrocarbon'? *(2 marks)*

ii) Alkenes such as ethene can be used to make polymers. Copy and complete the symbol equation below to describe how ethene can react to produce polythene. *(3 marks)*

n H∖C=C∕H → ⁄ ∖ H H

3 Many hydrocarbons are obtained from crude oil.
a) Describe how crude oil is formed. *(3 marks)*
b) Petrol and other important hydrocarbons can be separated from crude oil.
i) What is the name given to this separation process? *(1 mark)*
ii) Describe this process. *(3 marks)*
iii) Name two other chemicals that are obtained from crude oil by this technique. *(2 marks)*

4 Plastics are polymers and are usually non-biodegradable.
a) What is meant by the term 'polymer'? *(2 marks)*
b) What is meant by the term 'non-biodegradable'? *(2 marks)*
c) Name and describe the two main ways of disposing of polymers, giving a disadvantage of each. *(6 marks)*

5 Table 15.5 shows information about the energy released when some alkanes burn.

Table 15.5

Alkane	Number of carbon atoms	Energy released when burned/kJ mol^{-1}
methane	1	900
ethane	2	1550
propane	3	2200
butane	4	2850

a) Give the general formula of an alkane. *(1 mark)*
b) Pentane is an alkane with five carbon atoms.
i) Predict the energy released when pentane is burned. *(1 mark)*
ii) Suggest the structure of pentane. *(1 mark)*

16 Electrical circuits

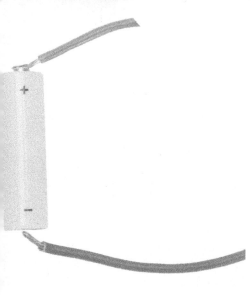

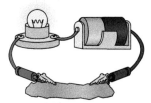

aluminium – conductor

iron – conductor

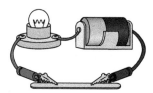

wood – insulator

▲ Figure 16.2 Testing if a material is a conductor

Specification points

This chapter covers sections 3.1.1 to 3.1.14 of the specification. It is about simple series and parallel circuits, the verification of Ohm's law, the definition of resistance and how to carry out an experimental investigation on which factors govern a wire's resistance.

Electricity

Electricity is an extremely versatile and useful form of energy. Many of our everyday activities depend on the use of electricity. It is hard to imagine life in today's society without it! Simple things such as entertainment, communications, transport and industry would simply grind to a halt if electricity ceased to exist. When a woollen jumper is taken off over a nylon shirt in the dark, you can hear crackles and see tiny blue electric sparks. The nylon shirt has become charged with static electricity.

We now know that there are two types of charge, positive charge and negative charge. The negative charge is due to the presence of electrons, and it is only these particles that we are concerned with in this chapter.

When we connect a battery across a lamp, the lamp lights up. We say that the connecting wire (copper) and the filament of the bulb (tungsten) are both electrical conductors. The plastic covering is not an electrical conductor. We say the plastic is an insulator.

How can we tell if a material is a conductor or an insulator? To do that we connect the material in a circuit containing a battery, a bulb and the material being tested. If the bulb lights up, it's a conductor. If the bulb does not light, it's an insulator.

You can see this test being used for aluminium, iron and wood in Figure 16.2.

▲ Figure 16.1 Without electricity, life would be unimaginable

In general, all metals are electrical conductors. Almost all non-metals are insulators, but there are a few exceptions. For example, graphite is a non-metal, but it conducts electricity. Some common conductors and insulators are shown in Table 16.1.

Table 16.1 Common conductors and insulators

Good conductors	Gold	Silver	Copper	Aluminium	Mercury	Platinum	Graphite
Insulators	Polythene	Rubber	Wool	Wax	Glass	Paper	Wood

Why are metals good conductors?

An electric current is a flow of electrically charged particles. At GCSE level, the charged particle involved is always the electron. Most electrons in atoms are bound by the positive nucleus to orbit in the surrounding shells. In metals, the outermost electron is often so weakly held that it can break away. We call such electrons free electrons. Some books call them delocalised electrons. Insulators contain no (or very few) free electrons.

So why does electrical current flow when we connect a metal wire between the terminals of a battery? An electric cell (commonly called a battery) can make electrons move, but only if there is a conductor connecting its two terminals to make a complete circuit. Chemical reactions inside the cell push electrons from the negative terminal round to the positive terminal. Figure 16.3 shows how an

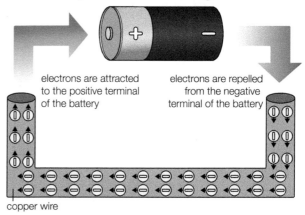

electrons are attracted to the positive terminal of the battery

electrons are repelled from the negative terminal of the battery

copper wire

▲ **Figure 16.3** The flow of electrons in a simple circuit

electric current would flow in a wire connected across a cell.

Electrons are repelled from the negative terminal of the cell because like charges repel, and are attracted to the positive terminal because unlike charges attract. This makes the electrons flow from the negative terminal to the positive terminal. Scientists in the nineteenth century thought that an electric current consisted of a flow of positive charge from the positive terminal of the cell to the negative terminal. This is now known to be incorrect.

To summarise:

▶ Electrical conductors, such as metals, have free electrons.
▶ Electrical insulators, such as non-metals, have no free electrons.
▶ Free electrons move from the negative to the positive terminal of the battery.
▶ Conventional current is said to flow from the positive terminal to the negative terminal of the battery.

Standard symbols

An electrical circuit may be represented by a circuit diagram with symbols for components. Circuit diagrams are easy to draw and are universally understood.

Table 16.2 Components and their symbols

Component	Symbol	Appearance
switch		
cell		
battery		
resistor		
variable resistor		
fuse		
voltmeter		
ammeter		
lamp		

Cell polarity

By convention, the long, thin line in the symbol for a cell is taken as the positive terminal.

The short, fat line is the negative terminal.

Cells can be joined together minus to plus to make a battery. This leaves a positive and a negative terminal free to be connected into a circuit. Cells connected in this way are said to be connected in series.

Portable stereo systems have a number of cells connected in series. The reason for this is that the system requires a large voltage to operate. Connecting cells in series to make a battery increases the voltage. For example, connecting four 1.5V cells in this way gives a 6V battery, as shown in Figure 16.4.

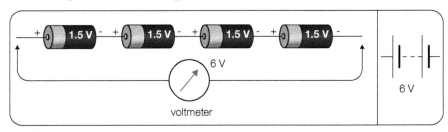

▲ **Figure 16.4** Cells correctly joined in series

You must be careful when connecting cells in series. If the polarity of one of the cells is reversed then the voltage is reduced dramatically. The voltages of two of the cells cancel each other out, leaving only one effective cell, as shown in Figure 16.5.

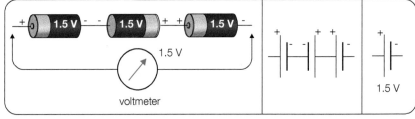

▲ **Figure 16.5** Cells incorrectly joined together

Notice that reversing the polarity of the cell in the middle has the effect of producing a battery of only 1.5 volts. The cells joined positive to positive cancel each other out.

Test yourself ✎

1 The cells in the diagram below are all identical. The total battery voltage is 1.6 V.

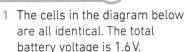

▲ **Figure 16.6**

   a) Calculate the voltage of each cell.
   b) Redraw the battery showing the connections which would give a battery capable of delivering the maximum possible voltage.
   c) State the maximum voltage that this battery could deliver.
2 The diagram shows a battery in series with a resistor (marked R) and a third component.

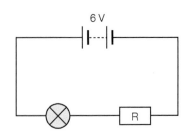

▲ **Figure 16.7**

   a) What is represented by the third component?
   b) Mark on the diagram with an arrow labeled E the direction in which electrons move in the circuit.

Prescribed practical P1: Ohm's law

This is the first experiment that Single Award Science students are required to do in preparation for the Unit 4 Practical Examination.

Prescribed practical P1 to show that the current and voltage are proportional for a metal wire at constant temperature.

Aims

- to pass an electric current through a wire
- to measure the current for different values of the voltage across the wire
- to take precautions to ensure the temperature of the wire is kept constant
- to plot a graph of voltage across the wire (y-axis) against current in the wire (x-axis)
- to use the graph to establish an equation linking voltage and current
- to determine the resistance of the wire

Apparatus

- low voltage power supply unit (PSU)
- rheostat
- ammeter
- voltmeter
- connecting leads
- resistance wire
- switch

Method

1. Prepare a table for your results as shown on page 160.
2. Ensure that the PSU is switched off and connect it to the mains socket.
3. Set up the circuit as shown in the circuit diagram. The device marked R represents the wire being tested.
4. Adjust the PSU to supply zero volts.
5. Switch on the PSU.
6. Record the voltage on the voltmeter and the corresponding current on the ammeter.
7. Switch off the PSU immediately after recording in the table values for voltage and current.
8. Wait for about two minutes to ensure the wire cools to room temperature.
9. Switch on the PSU and adjust the voltage (or the rheostat) so that the reading on the voltmeter increases by 0.5 V.
10. Repeat steps 6–9 until readings have been recorded for voltages ranging from zero to a maximum voltage of 6 V*. This is Trial 1.
11. Repeat the entire experiment to obtain a second set of current values. This is Trial 2.
12. Calculate the mean current from the two trials and enter the results in the table.
13. Plot the graph of voltage against mean current.

*It is necessary to ensure the wire's temperature remains constant (close to room temperature).

We do this by:

* keeping the voltage low (so that the current remains small)
* switching off the current between readings to allow the wire to cool.

Table for results

Table 16.3

Voltage/V	0.00	1.00	2.00	3.00	4.00	5.00	6.00
(Trial 1) Current/A							
(Trial 2) Current/A							
Mean current/A							
Ratio of voltage to current/Ω							

Circuit diagram

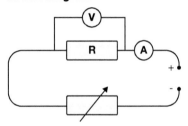

▲ **Figure 16.8**

Treatment of the results

Plot the graph of voltage in volts (vertical axis) against mean current in amps (horizontal axis) as shown in Figure 16.9.

Discussion of the results

The graph of V against I is a straight line through the origin. This tells us that the current in a metallic conductor is directly proportional to the voltage across its ends, provided the temperature remains constant.

This result is commonly called Ohm's law.

Measuring the resistance

The resistance of the wire does not change when the current and voltage change. The resistance of a wire at constant temperature depends only on three factors:

* the material from which the wire is made
* the length of the wire
* the area of cross section of the wire.

That is why the graph of V against I is a straight line through the origin. The ratio $V:I$ is constant throughout the experiment. It also means that, in this case, the slope of the graph of V against I is equal to the resistance of the wire. However, you should understand that measuring the slope of the V–I graph is not, in general, the correct way to measure resistance. This will become clearer. The next experiment is Prescribed practical P2.

$$V \quad = \quad I \quad \times \quad R$$
$$\text{voltage} \quad = \quad \text{current} \quad \times \quad \text{resistance}$$

The triangle in Figure 16.10 will help you to 'change the subject' in the resistance formula. To find the equation for current, put your thumb over I and it is clear that $I = \dfrac{V}{R}$

Similarly to find R, put your thumb over it and it is clear that $R = \dfrac{V}{I}$

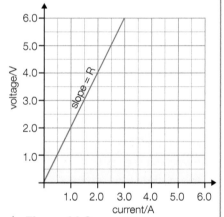

▲ **Figure 16.9**

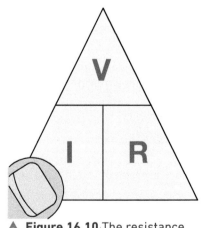

▲ **Figure 16.10** The resistance formula triangle

Tip

All mathematical questions in physics start with a formula, so it is essential that you learn your formulae if you are to do well in your exam!

Example

1 Find the missing values in the circuits below.

a)

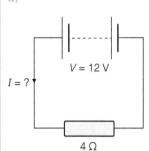

$I = ?$

$V = 12$ V

$4 \, \Omega$

b)   c)

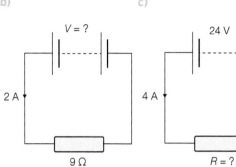

$V = ?$ 24 V

2 A   4 A

$9 \, \Omega$ $R = ?$

▲ **Figure 16.11**

Answer

a) $V = 12$V $R = 4\,\Omega$ $I = ?$

Current $I = \dfrac{V}{R}$

$= \dfrac{12}{4}$

$= 3$ A

b) $I = 2$A $R = 9\,\Omega$ $V = ?$

Voltage $V = I \times R$

$= 2 \times 9$

$= 18$V

c) $V = 24$V $I = 4$A $R = ?$

$R = \dfrac{V}{I}$

$= \dfrac{24}{4}$

$= 6\,\Omega$

3 Calculate the current flowing through a 10 Ω resistor which has a voltage of 20 V across it.
4 A resistor has a voltage of 15 V across it when a current of 3 A flows through it.
   Calculate the resistance of the resistor.
5 A current of 2 A flows through a 25 Ω resistor. Find the voltage across the resistor.
6 A voltage of 15 V is needed to make a current of 2.5 A flow through a wire.
   a) What is the resistance of the wire?
   b) What voltage is needed to make a current of 2.0 A flow through the wire?
7 There is a voltage of 6.0 V across the ends of a wire of resistance 12 Ω.
   a) What is the current in the wire?
   b) What voltage is needed to make a current of 1.5 A flow through it?
8 A resistor has a voltage of 6 V applied across it and the current flowing through it is 0.1 A. Calculate the resistance of the resistor.
9 A current of 0.6 A flows through a metal wire when the voltage across its ends is 3 V. What current flows through the same wire when the voltage across its ends is 2.5 V?

Show you can ?

1 Explain why it is necessary to keep the temperature of the resistance wire constant in the Ohm's law experiment, and describe how this is achieved.
2 Explain why it is necessary to obtain several values of current and voltage in the Ohm's law experiment, and describe how this is achieved.
3 Describe the characteristic V–I graph for a metal wire at constant temperature, and state what conclusion can be drawn from it.
4 State Ohm's law and the condition under which it is valid.

Series circuits

The total resistance of two or more resistors in series is simply the sum of the individual resistances of the resistors.

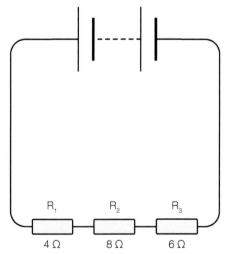

$$R_{total} = R_1 + R_2 + R_3$$

R_1 R_2 R_3
4 Ω   8 Ω   6 Ω

▲ **Figure 16.12** Calculating the total resistance of three resistors in a series circuit

In Figure 16.12, the three resistors could be replaced by a single resistor of 4 + 8 + 6 = 18 Ω.

Suppose in Figure 16.12, the battery voltage was 36V. By Ohm's law, the current in each resistor would be:

$$I = \frac{V}{R}$$
$$= \frac{36}{18}$$
$$= 2A$$

Applying Ohm's law to each resistor in turn:

The voltage across the 4Ω resistor would be 2 × 4 = 8V

The voltage across the 8Ω resistor would be 2 × 8 = 16V

The voltage across the 6Ω resistor would be 2 × 6 = 12V

The sum of the voltages across the resistors is 8 + 16 + 12 = 36V, which is exactly equal to the battery voltage.

The example has demonstrated some important facts about components in series:

▶ The current in each component in series is the same.

▶ The sum of the voltages across the separate components is equal to the voltage of the power supply.

Parallel circuits

Figure 16.13 below shows three resistors arranged in parallel across a 24V battery.

The voltage across each resistor is 24V.

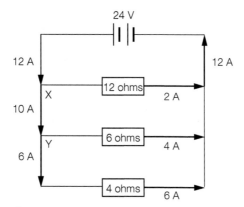

▲ **Figure 16.13** Three resistors in parallel

If we apply Ohm's law to each resistor in turn, we see that:

▶ the current in the 12Ω resistor is 24 ÷ 12 = 2A

▶ the current in the 6Ω resistor is 24 ÷ 6 = 4A

▶ the current in the 4Ω resistor is 24 ÷ 4 = 6A.

This can only happen if the current coming from the battery is 2A + 4A + 6A = 12A.

So 12A flow out of the battery (left side), and 12A return to the battery (right side).

Of the 12 A flowing out of the battery, 2 A flow through the 12 Ω resistor. The rest (12 − 2 = 10 A) carry on towards the 6 Ω resistor.

Of these 10 A, 4 A flow through the 6 Ω resistor. The rest (10 − 4 = 6 A) carry on towards the 4 Ω resistor.

This illustrates a very important idea. Look at the junction marked X. The current flowing into X is 12 A. Two currents flow away from X, 2 A and 10 A. Together they add up to 12 A, which is the same as the current flowing into X.

Similarly, the current flowing into junction Y (10 A) is the same as the sum of the currents flowing out of Y (6 A + 4 A).

We can summarise these findings for components in parallel:

▶ The voltage across each component is the same as that of the supply.
▶ The total current taken from the supply is the sum of the currents through the separate components.
▶ The sum of the currents entering a junction is always equal to the sum of the currents leaving it.

Test yourself

10 In Figure 16.14, the current in R_2 is 0.25 A.
  a) State the current in the other resistors.
  b) Calculate the voltage across each resistor.
11 Two resistors are connected in parallel, as shown in Figure 16.15.

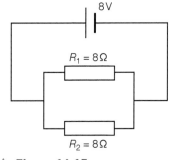

▲ **Figure 16.15**

  a) State the voltage across each of the resistors.
  b) Calculate the current in each resistor.
  c) Calculate the current being delivered by the battery.
12 a) For the circuit in Figure 16.16, calculate the current shown by ammeters A_1, A_2 and A_3.
  b) Calculate the voltage across each resistor.
13 a) In circuits A and B (Figure 16.17), all the lamps are identical. Copy and complete Table 16.4.

Table 16.4

Circuit	V_1 / V	V_2 / V	A_1 / A	A_2 / A
A	3	3	0.4	
B	3		0.1	

  b) Calculate the resistance of each lamp.

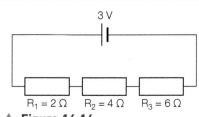

▲ **Figure 16.14**

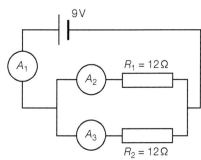

▲ **Figure 16.16**

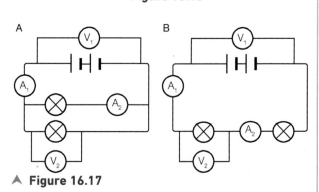

▲ **Figure 16.17**

Prescribed practical

Prescribed practical P2: Showing that when a metal wire is at a constant temperature, the resistance and length of wire are proportional

Aims

- to pass an electric current through a wire whose length can be varied
- to measure the current and voltage for different lengths of wire
- to take precautions to ensure the temperature of the wire is kept constant
- to calculate the resistance for each length of wire
- to plot a graph of resistance of the wire (*y*-axis) against length of the wire (*x*-axis)
- to use the graph to establish an equation linking resistance and length

Variables

- The independent variable is the length of the wire.
- The dependent variable is the resistance of the wire.
- The controlled variables are the temperature and the area of cross section of the wire.

Apparatus

- low voltage power supply unit (PSU)
- rheostat
- ammeter
- voltmeter
- connecting leads
- resistance wire
- switch
- metre ruler
- sticky tape

Method

1 Prepare a table for your results like the one shown in Table 16.5.

2 Measure and cut off one metre of nichrome resistance wire.

3 Attach the wire with sticky tape to a metre ruler – make sure there are no kinks in the wire.

4 Set up the circuit as shown in Figure 16.18.

5 Ensure that the PSU is switched off and connect it to the mains socket.

6 Adjust the PSU to supply about 1 V.

7 Connect the 'flying lead' so that the length of wire across the voltmeter is 10 cm.

8 Switch on the PSU.

9 Record the voltage on the voltmeter and the corresponding current on the ammeter.

10 Switch off the PSU immediately after recording in the table values for voltage and current.

11 Wait for about 2 minutes to ensure the wire cools to room temperature*.

12 Switch on the PSU again.

13 Repeat steps 7–12 until readings have been recorded for lengths of wire ranging from 10 cm to 90 cm.

14 Calculate the resistance of each length of wire, using $R = \dfrac{V}{I}$.

15 Plot the graph of resistance (y-axis) against length (x-axis).

It is necessary to ensure the wire's temperature remains constant (close to room temperature).

We do this by:

- keeping the voltage low (so that the current remains small)
- switching off the current between readings to allow the wire to cool.

Table for results

Table 16.5

Length of wire/cm	10	20	30	40	50	60	70	80	90
Voltage across wire/V									
Current in wire/A									
Resistance of wire/Ω									

Circuit diagram

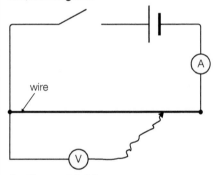

wire

▲ **Figure 16.18**

Treatment of the results

Plot the graph of resistance/Ω (vertical axis) against length/cm (horizontal axis).

Discussion of the results

The graph of resistance/Ω (vertical axis) against length/cm is a straight line through the origin. This tells us that the resistance of a metal wire is directly proportional to its length, provided the temperature and cross sectional area of the wire remain constant.

This means there is a mathematical relationship between the resistance, R, and the length, L.

The relationship is:

   $R = kL$

where k is the gradient of the graph.

Since $k = \dfrac{R}{L}$, the unit for k is Ω/cm (or Ω/m).

Note that the value of k depends on the material of the wire and its cross sectional area.

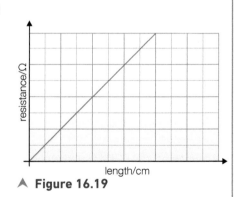

▲ **Figure 16.19**

Example

1 A reel of constantan wire of length 250 cm has a total resistance of 15.0 Ω. Calculate:

a) the resistance of 1.0 m of wire

b) the length of wire needed to have a resistance of 3 Ω

c) the resistance of a 90 cm length of the wire.

Answer

a) $k = \dfrac{R}{L}$

$= \dfrac{15\,\Omega}{250\,cm}$

$= 0.6\,\Omega/cm$

$= 60\,\Omega/m$, so the resistance of 1.0 m of wire is 60 Ω

b) $L = \dfrac{R}{k}$

$= \dfrac{3\,\Omega}{0.06\,\Omega/cm}$

$= 50\,cm$

c) $R = kL$

$= 6\,\Omega/m \times 0.9\,m$

$= 5.4\,\Omega$

Practical activity

H

You may be asked to answer questions on this material in Unit 3, but you will not be asked to carry out the experiment in Unit 4.

Resistance and cross sectional area

Aim

The aim is to set up an electrical circuit to allow you to:

- pass an electric current through wires of constant length but different cross sectional area
- measure the current for different values of the voltage and hence find the resistance
- measure the cross sectional area of each wire
- plot graphs of resistance (*y*-axis) against cross sectional area (*x*-axis) and resistance (*y*-axis) against 1/area
- determine the relationship between resistance and cross sectional area

Variables

- The dependent variable is the resistance of each wire.
- The independent variable is the cross sectional area of each wire.
- The controlled variables are the length of the wire, the temperature and the material from which the wire is made.

Apparatus

- low voltage power supply unit (PSU)
- rheostat
- ammeter
- voltmeter
- connecting leads
- filament lamp in a suitable holder
- switch
- wooden dowel or pencil or micrometer

▲ **Figure 16.20** Micrometer

Method

1 Prepare 7 samples of constantan wire, all 50 cm long and all of different cross sectional area.

2 Prepare a table for your results like the one shown in Table 16.6.

3 As a preliminary, use the micrometer screw gauge to measure the diameter, D, of one of the wires or, more simply, measure the length (l) of 20 turns of a resistance wire wound tightly together on a pencil or wooden dowel. Divide this length by 20 to calculate its diameter.

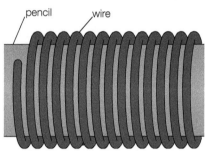

▲ **Figure 16.21** Wire wrapped around pencil

4 Calculate the cross sectional area, A, using $A = \dfrac{\pi D^2}{4}$ and record the data in the table of results.

5 Repeat this process for six further thicknesses of the same length of wire and same type of material.

6 Ensure that the PSU is switched off and connect it to the mains socket.

7 Set up the circuit as shown in the circuit diagram in Figure 16.22.

8 Switch on the PSU and adjust if necessary to obtain a voltage of 2 V.

9 Record the voltage on the voltmeter and the corresponding current on the ammeter.

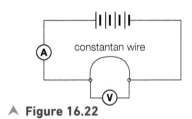

▲ **Figure 16.22**

10 Determine the resistance of this specimen of wire using $R = \dfrac{V}{I}$ and record the data in the table.

11 Repeat steps 7–10 for each of the other wire specimens.

12 Plot the graph of resistance against cross sectional area and resistance against 1/area.

Table for results

Table 16.6

Mean diameter/mm						
Cross sectional area/mm²						
Voltage/V						
Current/A						
Resistance/Ω						
1/area/1/mm²						

Treatment of the results

Plot the graphs of:

i) resistance/Ω against cross sectional area/mm²

ii) resistance/Ω against 1/area / 1/mm²

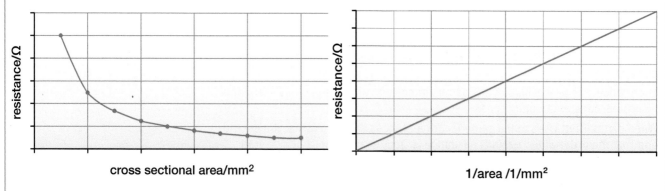

▲ **Figure 16.23**

Discussion of the results

The graph of R against A is a curve with decreasing gradient.

The graph of R against $1/A$ is a straight line through the origin. This tells us that the resistance is inversely proportional to the cross sectional area.

This means there is a mathematical relationship between the resistance, R and the cross sectional area, A.

The relationship is:

$R = \dfrac{k}{A}$

where k is the gradient of the straight line graph of R against $1/A$.

Since $k = RA$, the unit for k is Ωcm^2 (or Ωm^2 or Ωmm^2).

Note that, for a given wire, the value of k depends on the material of the wire and its length.

Example

1) A length of constantan wire has a resistance of 15.0 Ω and a cross sectional area of 0.3 mm². Find:

a) the resistance of the same length of constantan wire of area 0.9 mm².

b) the cross sectional area of the same length of constantan wire if its resistance is 9 Ω.

Answer

a) $k = RA = 15\ \Omega \times 0.3\ mm^2 = 4.5\ \Omega mm^2$

$R = \dfrac{k}{A} = \dfrac{4.5\ \Omega mm^2}{0.9\ mm^2} = 5\ \Omega$

b) $A = \dfrac{k}{R} = \dfrac{4.5\ \Omega mm^2}{9\ \Omega} = 0.5\ mm^2$

Practical activity

Investigating how the resistance of a metallic conductor at constant temperature depends on the material it is made from

The controlled variables in this investigation are the length and the thickness of wire.

It is reasonable to expect that the resistance should depend on the type of material from which a wire is made.

Using one metre of 32 swg copper wire, measure and record the resistance as before.

Then repeat the process using the same dimensions of wires such as manganin, nichrome, constantan and copper.

When comparing wires of the same length and cross sectional area, you should find that the order of increasing resistance is: copper, manganin, constantan and nichrome.

Show you can

1 Describe how the resistance of a piece of metal wire depends on:
   a) its length
   b) its cross-sectional area
   c) the material from which it is made.

2 Describe in detail the experiments which prove how the resistance of a piece of metal wire depends on:
   a) its length
   b) its cross-sectional area
   c) the material from which it is made.

Test yourself

14 A school buys a reel of constantan wire. The supplier's data sheet says that the wire has a resistance of 2.5 Ω/m. Calculate:
   a) the length of wire a technician must cut from the reel to give a resistance of 2 Ω
   b) the resistance of a 120 cm length of wire cut from the reel.

15 When a current of 0.15 A flows through a 48 cm length of eureka wire the voltage across its ends is 0.90 V. What length of the same type of wire would give a current of 0.36 A when the voltage across its ends is 1.44 V?

16 An 80 cm length of wire A has a resistance of 2.40 Ω. The resistance of a 50 cm length of wire B is 1.2 Ω. A student cuts a 30 cm length, L_1, of wire from a reel of wire A and a 40 cm length, L_2, from a reel of wire B.
   a) Which length, L_1 or L_2, has the greater resistance? Explain your reasoning.
   b) Sketch a graph of resistance against length for wire A and wire B, and state which graph has the larger gradient.

17 A technician cuts an 80 cm length of wire from a reel marked 3.0 Ω/m. The technician joins the two free ends of the wire together to form a loop. She then attaches two crocodile clips to the wire at opposite ends of a diameter.
   a) Explain why the total resistance between the crocodile clips is 0.6 Ω.
   b) In what way, if at all, does the total resistance between the crocodile clips change, if one of the clips is moved along the wire towards the other. Explain your reasoning.

18 A 50 cm length of wire with a diameter of 0.2 mm has a resistance of 1.6 Ω. Find the resistance of:
   a) a 75 cm length of wire of the same material and same diameter
   b) a 50 cm length of wire of the same material and diameter 0.4 mm
   c) a 75 cm length of wire of the same material and diameter 0.4 mm.

19 A piece of wire 20 cm long has a diameter of 0.3 mm and a resistance of 0.8 Ω.
   Show that the resistance of a wire of the same material that has a length of 80 cm and diameter 0.6 mm is also 0.8 Ω.

Practice questions

1 a) What flows in the direction indicated by the arrow in Figure 16.24? *(1 mark)*

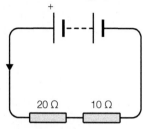

Figure 16.24

b) Copy the circuit diagram and mark on it an arrow to show the direction in which charged particles flow through the two resistors. *(1 mark)*

c) What name is given to these charged particles? *(1 mark)*

d) The current in the larger resistor is 0.6 A.
  i) State the size of the current in the smaller resistor. *(1 mark)*
  ii) Show that the battery voltage is 18 V. *(3 marks)*

2 a) Two identical resistors are connected in parallel across a battery. Their combined resistance is 12 Ω. What is the resistance of each resistor? *(1 mark)*

b) A different pair of identical resistors is connected in series across a battery. Their combined resistance is also 12 Ω. What is the resistance of each resistor? *(1 mark)*

3 In Figure 16.25 A and B, the lamps are identical and each has an internal resistance of 6 Ω. Each cell in the battery can supply a voltage of 1.5 V. For each circuit, find the reading which would be shown on the four meters. *(4 marks)*

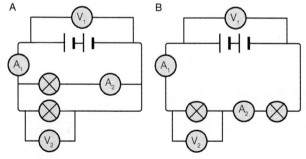

Figure 16.25

4 Four identical resistors are arranged as shown in Figure 16.26.

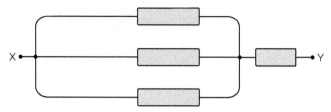

Figure 16.26

The current entering at X is 3 mA, and the voltage between X and Y is 12 mV. Calculate:

a) the total resistance between X and Y *(3 marks)*
b) the current in each resistor *(2 marks)*
c) the voltage across each resistor *(2 marks)*
d) the resistance of each resistor. *(2 marks)*

5 In the circuit diagram shown in Figure 16.27, resistors R_1 and R_2 have resistances of 40 Ω and 20 Ω respectively.

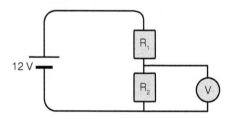

Figure 16.27

a) Calculate the voltage you would expect to observe on the voltmeter. *(3 marks)*
b) What assumption have you made about the resistance of the voltmeter itself? *(1 mark)*

6 A student investigating the relationship between the resistance and the area of cross section of a resistance wire of length 1.00 m obtains the following results.

a) Copy Table 16.7 and enter the missing data. *(3 marks)*

Table 16.7

Area of cross section of wire/mm²	0.5	1.0	2.0	3.0	4.0
Resistance/Ω	24.0		6.0		

b) Calculate the resistance of 50 cm of a wire of cross sectional area 1.5 mm² made from the same material. *(3 marks)*

17 Household electricity

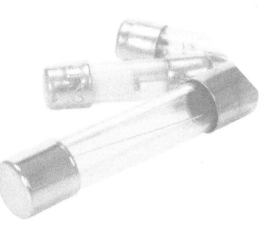

The power equation

When an electric current flows through a resistance, heat energy is generally produced. The amount of heat produced per second (or the electrical power, P) is given by the equation:

electrical power = voltage × current

This is often expressed by the equation $P = I \times V$ and is commonly called Joule's law of heating.

In this equation, P is measured in watts (W) or joules per second (J/s), I is measured in amps (A) and V is measured in volts (V).

Domestic appliances such as toasters, hairdryers and TVs have a power rating marked on them in watts or in kilowatts (1 kW = 1000 W).

Example

1 If 0.5 A flows through a bulb connected across a 6 V power supply for 10 seconds, calculate a) the electrical power and b) the amount of electrical energy transferred.

Answer

a) $P = I \times V$

$= 6 \times 0.5$

$= 3 \text{ W or } 3 \text{ J/s}$

b) energy = power × time

$= 3 \times 10$

$= 30 \text{ J}$

2 A study lamp is rated at 60 W, 240 V. How much current flows in the bulb?

Answer

$P = V \times I$

$60 = 240 \times I$

$I = \dfrac{60}{240}$

$= 0.25 \text{ A}$

Test yourself

1 How much electrical energy does a 1000 W convector heater consume in one hour?
2 In 10 seconds, an electric toaster consumes 15 000 joules of energy from the mains supply. What is its power:
   a) in watts
   b) in kilowatts?
3 A study lamp draws a current of 0.25 A at 240 V from the mains supply. Calculate:
   a) the power
   b) the amount of energy it consumes in 60 seconds.
4 The starter motor of a car has a power rating of 960 W.
   a) If it is switched on for 5 seconds, how much energy does it use?
   b) The same starter motor is powered by connecting it to a 12 V car battery. How much current does it use?

Fuses

A fuse is a device that is meant to prevent damage to an appliance.

The most commonly used fuses are either a 3A (red) fuse for appliances up to 720W, or a 13A (brown) fuse for appliances between 720W and 3kW.

If a larger-than-usual current flows, the fuse wire will melt and break the circuit.

Selecting a fuse

Every appliance has a power rating. How much current the appliance will use is found using the power formula:

$$power = voltage \times current$$

For example, a jig-saw has a power of 350W. The current it draws when connected to the mains is given by:

$$current = \frac{power}{voltage}$$
$$= \frac{350}{240}$$
$$= 1.46\,A$$

This is the normal current the device uses. A larger current could destroy it.

A 3A fuse would allow a normal working current to flow and protect the jig-saw from larger currents. A 13A fuse would allow a dangerously high current to flow without breaking the circuit. It is important to use the correct size of fuse.

Remember that a fuse protects the appliance. It does not protect the person using the appliance. It can take 1 to 2 seconds for a fuse wire to melt – enough time for the user to receive a fatal electric shock.

▲ **Figure 17.1** Fuses

If the fuse in the plug is less than it should be ...	the fuse will immediately blow when the appliance is switched on.
If the fuse in the plug is more than it should be ...	the fuse will probably not blow, even when a dangerously high current is flowing, so the appliance is not protected!

We will see later how the fuse and the earth wire in the plug work together to protect the user from electric shock.

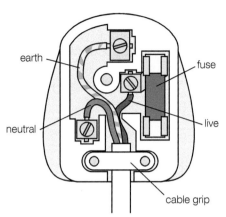

Figure 17.2 A three–pin plug

Show you can ?

1 State Joule's law as an equation.
2 Explain the consequences of using an incorrect fuse in a plug.
3 Describe the wiring in a three–pin plug.

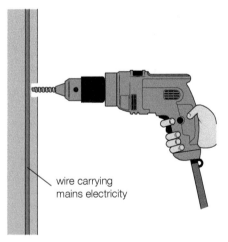

Figure 17.3 If there is no earth wire connected to the casing of the drill, the current will flow through the person

How to wire a three-pin plug

The wire with the blue insulation is the neutral wire – connect this to the left-hand pin.

The brown insulated wire is the live wire – connect this to the right-hand pin.

The wire with the yellow and green insulation is the earth wire – connect this to the top pin.

Each of these wires should be wrapped around its securing screw so that it is tightened as the screw is turned.

Insert the correct cartridge fuse into its holder beside the live wire.

Finally, fix the 3-core cable tightly with the cable grip and screw on the plug-top.

Each pin in the plug fits into a corresponding hole in the socket. The earth pin is longer than the others so that it goes into the socket first and pushes aside safety covers, which cover the rear of the neutral and live holes in the socket. Figure 17.2 shows the layout of a three-pin plug.

What the live, neutral and earth wires do

When a mains appliance is switched on, the current in the live and neutral wires is always the same. However, the live wire is the dangerous wire – it always carries a high voltage (around 230V) with respect to the neutral wire.

The neutral wire is around 0V, but this does not mean it is safe to touch!

Always respect mains electricity. According to the Health and Safety Executive, 144 employees died at work in the UK as a result of mains electricity in 2015. Another 67 people were electrocuted outside their place of work.

Normally, no current at all flows in the earth wire. This wire is connected to a plate that is buried deep under the earth.

How do the earth wire and fuse work together to prevent harm to the user?

Suppose a fault develops in an electric fire, and the heating element comes into contact with the metal casing of the fire. The casing would become live, and if someone were to touch it they could get a fatal electric shock as the current rushed through their body to earth. The earth wire prevents this – it offers a low resistance route of escape, enabling the current to go to earth by a wire rather than through a human body. Because of this low resistance, the current though the fuse will be large. After a few seconds, the large current will melt the fuse.

Together, the fuse and the earth wire protect the user from electric shock. The person in Figure 17.3 will be protected if the drill makes contact with the mains wire as long as there is an earth wire connected to the casing of the drill.

Any appliance with a metal casing could become live if a fault developed, so such appliances nearly always have a fuse fitted in the plug.

The switch and fuse must be placed on the live side of the appliance (Figure 17.4). Why is this important?

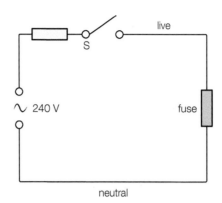

▲ **Figure 17.4**

If a fault occurs and the fuse blows, the live, dangerous wire is disconnected. If the fuse was on the neutral side, the appliance would still be live, even when the fuse had blown.

Switches are also placed on the live side for the same reason. If the switch was on the neutral side, the appliance would still be live, even when the switch was in the OFF position.

Double insulation

Appliances such as vacuum cleaners and hairdryers are usually double insulated. The appliance is encased in an insulating plastic case and is connected to the supply by a two-core insulated cable containing only a live and a neutral wire. Any metal attachments that a user might touch are fitted into the plastic case so that users do not make a direct connection with the motor or other internal electrical parts. The symbol for double insulated appliances is shown in Figure 17.5.

▲ **Figure 17.5**

Show you can

1 Explain the function of the live, neutral and earth wires.
2 Explain why the fuse and switch in a mains circuit should always be on the live side of the power supply.
3 Explain how the fuse can protect an appliance.
4 Explain how the fuse and the earth wire together can protect a user from electric shock.
5 Explain what is meant by double insulation and describe how it protects the user.

H

Paying for electricity

Electricity companies bill customers for electrical energy in units known as kilowatt-hours (kWh). These are sometimes called 'units' of electricity.

One kilowatt-hour is the amount of energy transferred when 1000 W is delivered for one hour. You should be able to prove for yourself that:

1 kWh = 3 600 000 J = 3.6 MJ

The following two formulae are very useful in calculating the cost of using a particular appliance for a given amount of time:

number of units used = power rating (in kilowatts) × time (in hours)

total cost = number of units used × cost per unit

There are two important numbers on an electricity bill, the present meter reading and the previous meter reading, shown in the first two columns of Figure 17.6.

Tip

Remember when calculating the number of units of electricity used the power must be in kW – so if necessary divide by 1000 to change watts to kW – and the time must be in hours – so if necessary divide any time in seconds by 3600.

Northern Electricity Board Customer account no: 3427 364

Present meter reading	Previous meter reading	Units used	Cost per unit (incl. VAT)	£
57139	55652	1487	15.0p	£223.05

Figure 17.6 A typical electricity bill

The difference between the current reading and the previous reading is the number of units used.

In this particular example, 57 139 – 55 652 = 1487 units (kWh).

1487 units have been used.

If the cost of a unit is known, then the total cost of the electricity used can be determined.

In this particular example: 1487 units at 15.0 pence per unit = 22 305 pence or £223.05.

Example

1 a) Find the number of units used by a 2800 W oven when it is switched on for five hours.

b) Calculate the cost of using this oven for that time if electricity costs 16 pence per kilowatt hour.

Answer

a) number of units used = power rating (in kilowatts) × time (in hours)

= 2.8 kW × 5 h

= 14 kWh

b) total cost = number of units used × cost per unit

= 14 kWh × 16 pence

= 224 pence

= £2.24

Show you can

1 Explain what is meant by a kilowatt-hour (kWh).
2 Use the power rating of an appliance to calculate the energy it uses in a given time.
3 Calculate the cost of using a given appliance for a given time.
4 Use meter readings to calculate the cost of household electricity.

Practice questions

1 An electric fire is connected to the mains supply by means of a three-pin plug.
   a) The electric fire has a rating of 2000 W when used on a 230 V mains supply. Calculate which fuse would be most suitable for use with this fire: 1 A, 3 A, 5 A or 13 A. *(4 marks)*
   b) Describe the size of the current in the live, neutral and earth wires when the fire is switched on and working properly. *(1 mark)*
   c) The live wire becomes loose and comes into contact with the metal body of the electric fire. Describe the danger that could arise when the electric fire is switched on. *(1 mark)*
   d) How should the earth wire be connected to remove this danger? *(1 mark)*
   e) How should the fuse be connected so as to remove this danger? *(1 mark)*
   f) Explain how the action of the earth wire and the fuse remove this danger. *(1 mark)*

2 Mrs Johnston's electricity meter was read at monthly intervals.
   Reading on 1 April 2017   11 897 kWh
   Reading on 1 May 2017   12 107 kWh
   a) How many units of electricity were used in the Johnston home during April? *(1 mark)*
   b) If 1 unit of electricity costs 15p, calculate the cost of electricity to Mrs Johnston during April. Show clearly how you get your answer. *(2 marks)*

3 An electric fire is wired using a three-pin plug as shown in Figure 17.7.

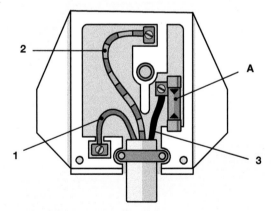

Figure 17.7

   a) Name the part labelled A. *(1 mark)*
   b) Which of the wires 1, 2 or 3 should be connected to the metal casing of the fire? *(1 mark)*
   c) State the colours of the insulation on wires 1, 2 and 3. *(3 marks)*

4 A television set is marked 240 V, 80 W.
   a) Explain carefully what these numbers mean. *(2 marks)*
   b) The flex that connects a brand new television to the mains has only two wires inside it. An electrician confirms that there should only be two wires inside the plug. Explain why only two wires are needed. *(3 marks)*
   c) To which of these wires should the switch on the television be connected? *(1 mark)*
   d) Apart from allowing the user to switch the television on and off, this is done for another reason. What is this other reason? *(1 mark)*
   e) Explain how the owner of this television is protected from possible electric shock. *(3 marks)*

5 An electric oven is rated at 8 kW.
   a) Calculate the cost of using the oven to cook for 2 hours. The cost of electricity is 15p per unit. Show clearly how you get your answer. *(3 marks)*
   b) When the oven is on, the same current passes through both the cable and the heating elements. Explain why the cable does not heat up. *(2 marks)*

6 Calculate how much electrical energy, in kilowatt-hours, is used to power:
   a) a 100 W lamp on for 12 hours *(2 marks)*
   b) a 250 W television on for 4 hours *(2 marks)*
   c) a 2400 W kettle on for 5 minutes? *(2 marks)*

7 An electric shower is rated at 230 V, 15 A.
   a) Calculate the electrical power used by the shower heater. *(2 marks)*
   b) Calculate the cost of a 10-minute shower if 1 kWh costs 12p. *(2 marks)*

8 a) Calculate the amount of electrical energy, in joules, used by a 1000 W electric fire in 1 hour. *(3 marks)*
   b) What common name is given to this quantity of energy? *(1 mark)*

9 Cartridge fuses are normally available in 3 A, 5 A or 13 A.
   a) What could happen if you used a 3 A fuse in the plug for a 3 kW electric fire? *(1 mark)*
   b) Why is it bad practice to use a 13 A fuse in a plug for a 60 W study lamp? *(1 mark)*

c) What size of fuse would you use for a hairdryer labelled 230V, 800W? Explain how you worked out your answer. *(2 marks)*

10 What is the highest number of 60W bulbs that can be run off the 230V mains if you are not going to overload a 5A fuse? *(3 marks)*

11 A 13A socket is designed to allow a current of 13A to be drawn safely from it. Mr White connected the following appliances to a single 13A socket using a 4-way extension lead: a 2.4kW electric kettle; a 3kW dishwasher; an 800W television; a 1300W toaster.

a) Calculate the current through each appliance, assuming that the supply voltage is 230V. *(4 marks)*

b) Assuming that the plug from the extension lead contained a 13A fuse, what would happen if he attempted to use all the appliances at the same time? *(1 mark)*

18 Energy

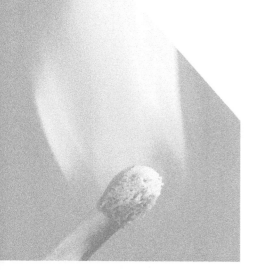

Specification points

This chapter covers sections 3.3.1 to 3.3.11 of the specification. It includes the definition of efficiency and the application of the formula, the principle of conservation of energy, common energy transducers, and kinetic and gravitational potential energy.

Forms of energy

Energy comes in many different forms. The most common are chemical, heat, electrical, sound, light, magnetic, strain energy, kinetic and gravitational potential energy. Table 18.1 summarises some of the main energy forms.

Table 18.1 Forms of energy

Energy form	Definition	Examples
chemical	the energy stored in a substance that is released on burning	coal, oil, natural gas, peat, food
heat	the energy form which gives the feeling of warmth	fire
electrical	energy produced by a battery or an electrical generator	batteries
sound	energy which comes from vibrations in the air	what we hear coming from a musical instrument
light	the energy responsible for photosynthesis and which enables us to see	radiation from the Sun
magnetic	the energy stored around a magnet or in a magnetic field	the energy around a bar magnet which causes it to attract a steel paper clip
strain	the energy stored in an object when it is stretched or compressed	the energy stored in a stretched catapult or the compressed sheet of a trampoline
kinetic	the energy of a moving object	a moving train
gravitational potential	energy possessed by a mass because of its height above the Earth's surface	the energy stored in a mass because of its height above the Earth

All energy is measured in joules (J). The energy forms listed above are not the only energy forms that exist. We could have discussed nuclear energy or wave energy or tidal energy and so on.

One of the fundamental laws of physics is the principle of conservation of energy.

This states that:

> Energy can be changed from one form to another, but the total amount of energy does not change.

Some people prefer to state the principle of conservation of energy slightly differently:

> Energy can neither be created nor destroyed; it can only change its form.

We can show energy changes in an energy flow diagram (Figure 18.1).

What energy changes take place when we strike a match?

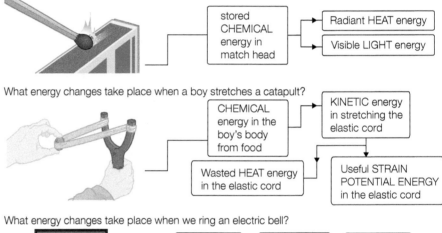

What energy changes take place when a boy stretches a catapult?

What energy changes take place when we ring an electric bell?

▲ **Figure 18.1** Energy changes

Energy flow diagrams are a useful way of showing the main energy changes taking place, but they have two major limitations. They usually do not show:

▶ all the energy transformations taking place
▶ the amount of energy being changed from one form to another.

Test yourself

1 List eight different types of energy and give an example of each.
2 State the principle of conservation of energy.

Kinetic energy

The kinetic energy of an object is the energy it has because it is moving. It can be shown that an object's kinetic energy is given by the formula:

$$\text{kinetic energy} = \frac{1}{2} \times \text{mass} \times \text{velocity}^2 \text{ or } KE = \frac{1}{2}mv^2$$

where m is the mass in kg and v is the velocity of the object in m/s.

Example

1 A car of mass 800 kg is travelling at 15 m/s. Find its kinetic energy.

Answer

$$KE = \frac{1}{2}mv^2$$
$$= \frac{1}{2} \times 800 \times 15^2$$
$$= 90\,000 \text{ J}$$

2 A bullet has a mass of 20 g and is travelling at 300 m/s. Find its kinetic energy.

Answer

$$20\,g = 0.02 \text{ kg}$$
$$KE = \frac{1}{2}mv^2$$
$$= \frac{1}{2} \times 0.02 \times 300^2$$
$$= 900 \text{ J}$$

3 Find the speed of a boat if its mass is 1200 kg and it has a kinetic energy of 9600 J.

Answer

$$KE = \frac{1}{2}mv^2$$
$$9600 = \frac{1}{2} \times 1200 \times v^2$$
$$v^2 = \frac{9600}{600}$$
$$v = 4 \text{ m/s}$$

4 Find the mass of a snooker ball if it has a KE of 0.48 J when moving at 4 m/s.

Answer

$$KE = \frac{1}{2}mv^2$$
$$0.48 = \frac{1}{2} \times m \times 4^2$$
$$0.48 = 8 \times m$$
$$m = 0.06 \text{ kg}$$
$$= 60 \text{ g}$$

Gravitational potential energy

When any object with mass is lifted, work is done on it against the force of gravity. The greater the mass of the object and the higher it is lifted, the more work has to be done. The work that is done is only possible because some energy has been used. This energy is stored in the object as gravitational potential energy.

$$GPE = mgh$$

where m is the mass in kg, g is the gravitational field strength in N/kg and h is the vertical height in m.

It is important to remember that 1 kg has a weight of 10 N on Earth. This is just another way of saying that the gravitational field strength, g, on Earth is 10 N/kg.

Example

1 Find the gravitational potential energy of a mass of 500 g when raised to a height of 240 cm. Take $g = 10$ N/kg.

Answer

$$500\,g = 0.5\,kg$$
$$240\,cm = 2.4\,m$$
$$GPE = mgh$$
$$= 0.5 \times 10 \times 2.4$$
$$= 12\,J$$

2 How much heat and sound energy is produced when a mass of 1.2 kg falls to the ground from a height of 5 m? Take $g = 10$ N/kg.

Answer

Heat and sound energy produced = original GPE
$$GPE = mgh$$
$$= 1.2 \times 10 \times 5$$
$$= 60\,J$$

3 What mass of water is held in the reservoir of a hydroelectric power station if it stores 4 000 000 000 J of gravitational potential energy and the water is at an average height of 80 m above the turbines?

Answer

$$GPE = mgh$$
$$4\,000\,000\,000 = m \times 10 \times 80$$
$$= 800 \times m$$
$$m = 4\,000\,000\,000 \div 800$$
$$= 5\,000\,000\,kg$$

4 A marble of mass 50 g falls to the Earth. At the moment of impact, its kinetic energy is 1 J. From what height did it fall?

Answer

$$50\,g = 0.05\,kg$$
KE at impact = GPE at start
$$1 = mgh$$
$$= 0.05 \times 10 \times h$$
$$= 0.5 \times h$$
$$h = 2\,m$$

Renewable and non-renewable energy

Energy can be classified as renewable or non-renewable. Renewable energy is defined as energy that is collected from resources that will never run out or that are naturally replenished on a human timescale.

Examples of renewable energy are sunlight, wind, rain, tidal, waves, wood and geothermal heat. Biomass is renewable if the plant material used is re-grown, but it is considered non-renewable if the plants used are not re-grown.

Non-renewable energy sources are those that will eventually run out at some time in the future. There is a finite supply of non-renewable energy resources.

Examples of non-renewable energy resources are fossil fuels such as coal, oil and natural gas. Fossil fuels all come from the dead remains of plants and animals subject to high temperatures and pressures in the Earth's crust. They are considered non-renewable because their formation takes millions of years.

Nuclear energy based on fission is also non-renewable since it relies on supplies of uranium and plutonium which will eventually run out. Table 18.2 lists examples of renewable and non–renewable energy resources.

Table 18.2 A summary of renewable and non-renewable energy resources

Energy resource	Renewable (R) or non-renewable (N)	Comment
fossil fuels such as coal, oil, natural gas, lignite, peat	N	coal might last another 300 years at current rates of use, the other fossil fuels will run out sometime in the 21st century. coal is a dirty fuel and produces the most greenhouse gases per unit of electricity generated.
nuclear fuels such as uranium and plutonium	N	supporters say it will last much longer than most fossil fuels and produces no greenhouse gases.
wind farms	R	they are unreliable because we cannot depend on the wind.
waves	R	they are unreliable and are still the subject of much research.
tides	R	unlike the wind and waves, the tides are reliable. but tidal barrages to produce electricity are very expensive. they can destroy plant and animal habitats, and it is difficult to find appropriate sites that would not be too affected.
geothermal (heat)	R	the heat energy is extracted from hot rocks deep underground. it is renewable, because the rocks are continually being re-heated by processes deep under the earth's crust, but there are limited suitable sites available.
biomass such as wood, grass (turned into biogas)	R (if plants re-grown) N (if plants not re-grown)	these are becoming more popular and are (almost) carbon neutral, so they make (almost) no net contribution to climate change.

Efficiency

Energy is wasted in every physical process. Consider, for example, a car engine. The input energy is chemical energy in the form of petrol. The useful output energy is the kinetic energy of the car. But the engine also produces heat and sound, both of which are wasted energy forms.

1 An electric kettle uses 400 000 J of electrical energy to boil some water. In the process, 360 000 J of heat energy passes into the water. Calculate the kettle's efficiency.

Answer

useful energy output
(passed into water) = 360 000 J
total energy input = 400 000 J

efficiency = $\dfrac{\text{useful energy output}}{\text{energy input}}$

$= \dfrac{360\,000}{400\,000}$

$= 0.9$

Therefore:
- 90% of the electrical energy is used to boil the water
- 10% of the energy supplied is wasted

A small amount of heat will be lost as some water evaporates. If the kettle whistles, a small amount of energy will also be lost as sound.

2 A lift motor has an efficiency of 0.45. The motor takes in 16 000 J of electrical energy. Calculate the useful energy output.

Answer

efficiency = $\dfrac{\text{useful energy output}}{\text{energy input}}$

$0.45 = \dfrac{\text{useful energy output}}{16\,000}$

useful energy = 7200 J
output

Efficiency is a physical quantity that measures the proportion of the input energy which appears as useful output energy. Efficiency is defined by the formula:

$$\text{efficiency} = \dfrac{\text{useful energy output}}{\text{energy input}}$$

As efficiency is a ratio, it has no units. The useful energy output will always be less than the total energy input, so the efficiency of a machine is always less than 1.

3 a) State the difference between renewable and non-renewable energy resources and give two examples of each.
 b) Under what condition is biomass considered a renewable energy resource?
4 a) What are fossil fuels?
 b) Give the names of three common fossil fuels.
 c) Give a reason why governments are keen to find replacements for fossil fuels.
5 An oil tanker has a mass of 100 000 tonnes. Its kinetic energy is 200 MJ. Calculate its speed. (1 tonne = 1000 kg; 1 MJ = 1 000 000 J)
6 A ball of mass 2 kg falls from rest from a height of 5 m above the ground. Copy the table below and complete it to show the gravitational potential energy, the kinetic energy, the speed and the total energy of the falling ball at different heights above the surface.
Table 18.3

Height above ground/m	Gravitational PE/J	Kinetic energy/J	Total energy/J	Speed/m/s
5.0		0	100	0
4.0				4.47
	64			
1.8		64		
0.0				

7 A car of mass 800 kg is travelling at a steady speed. The kinetic energy of the car is 160 000 J. Show that the speed of the car is 72 km per hour.
8 Figure 18.2 shows energy transfers in a mobile phone.

25 J input electrical energy
useful sound energy
20 J wasted energy

▲ **Figure 18.2**
 a) Use the figures on the diagram to calculate the phone's efficiency.
 b) What principle of physics did you use to calculate the useful sound energy produced?

1 Copy and complete Table 18.4 to show the main energy change that each device or situation is designed to bring about. The first one has been done for you.

Table 18.4

Device / situation	Input energy form		Useful output energy form
Microphone	sound energy	→	electrical energy
Electric smoothing iron	.... energy	→	.... energy
Loudspeaker	.... energy	→	.... energy
Coal burning in an open fire	.... energy	→	.... energy
A candle flame	.... energy	→	.... energy and .... energy

(4 marks)

2 Which of these are energy forms?

sound pressure force weight electricity heat

(3 marks)

3 Electricity companies say that electricity is a 'clean' fuel. Why is this statement misleading?

(2 marks)

4 Name an energy resource that does not depend directly on energy from the Sun. *(1 mark)*

5 Labels like this are used to advise customers about the efficiency of domestic appliances.

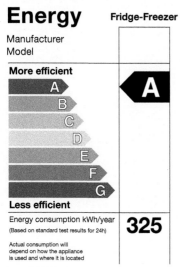

Figure 18.3

Mrs Johnston wants to buy a new dishwasher. She has a choice of a Grade D dishwasher or a Grade A dishwasher.
a) Which would cost less to run? *(1 mark)*
b) Mrs Johnston selects a dishwasher. The manufacturer's datasheet informs her that for every 1000 J of electrical energy input, only 120 J of energy is wasted.
Calculate the efficiency of this dishwasher.

(3 marks)

c) Copy and complete Figure 18.4 to describe the main energy changes that take place in a dishwasher. *(4 marks)*

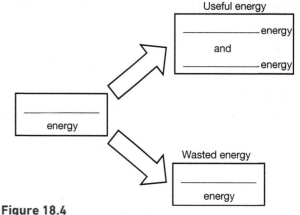

Figure 18.4

6 The pie chart shows the main sources of the energy used in the world today.

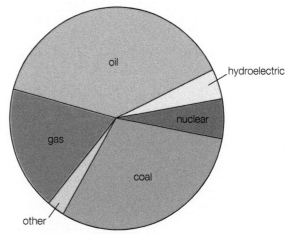

Figure 18.5
a) On which two fuels are we most dependent?

(2 marks)

b) Write down an energy resource in the diagram which is a fossil fuel. *(1 mark)*
c) In what way is this pie chart likely to be different at the end of the 21st century? *(1 mark)*

7 A concrete block of mass 20kg is used to drive a wooden stake into the ground. The block is first raised to a height of 3.2 m above the top of the stake, as shown in Figure 18.6a.

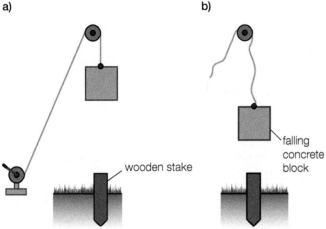

a)   b)

wooden stake

falling concrete block

Figure 18.6

a) Calculate the gravitational potential energy of the block relative to the top of the stake. *(3 marks)*

b) Using the principle of conservation of energy, calculate the speed of the concrete block as it hits the stake. *(4 marks)*

8 Figure 18.7 shows a large pendulum which has been pulled to one side and held at position A. The dotted line shows the position of the pendulum as it swings through the mid point, position B.

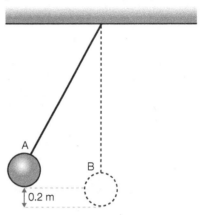

A

B

0.2 m

Figure 18.7

a) Copy and complete the following sentence.

As the pendulum swings from position A to position B, ... energy changes to .... energy. *(2 marks)*

The pendulum, which has a mass of 0.75 kg, is released from point A.

b) Use the principle of conservation of energy to calculate the speed of the pendulum as it passes through point B. *(4 marks)*

19 Electricity generation

Electromagnetic induction

In 1831, Michael Faraday discovered that electricity can be made without using a battery. The only apparatus required is a coil of wire and a magnet.

As shown in Figure 19.1, when the magnet moves in and out of the coil, a current is generated. It is only when the magnet is actually moving that electricity is produced. The ammeter shown in Figure 19.1 allows us to measure how much current is flowing.

This process is called electromagnetic induction. It is the basic principle behind the generation of electricity in dynamos (Figure 19.2) and in power-station generators. Electromagnetic induction is the process of obtaining a voltage by changing the magnetic field near a conductor.

We can induce more electric current by:

▶ using a stronger magnet
▶ using a coil with more turns
▶ moving the magnet faster.

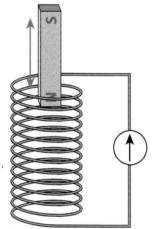

▲ **Figure 19.1** Generating electricity with a magnet and a coil of wire

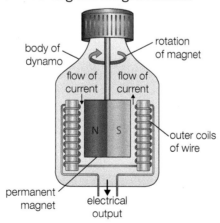

▲ **Figure 19.2** In a bicycle dynamo, the magnet turns inside a coil of wire to produce electricity

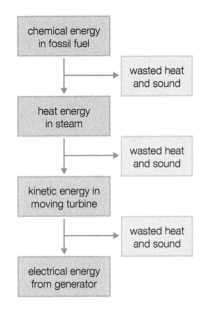

▲ **Figure 19.3** The energy flow diagram in a fossil fuel power station

Power stations

Most power stations use the principle of electromagnetic induction (moving a magnet relative to a coil of wire) to make electricity.

Figure 19.4 shows what happens inside a typical fossil fuel (coal, oil or gas) power station.

Inside the power station, there are four main stages:

▶ The fuel is burned, releasing heat.

▶ The heat is used to turn water into steam.

▶ The steam is used to drive a turbine, which is connected to a generator.

▶ The generator generates electricity when the magnet moves relative to a coil of wire.

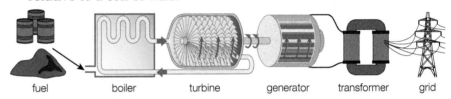

fuel   boiler   turbine   generator   transformer   grid

▲ **Figure 19.4** What happens inside a fossil fuel power station

Transformers

Power stations typically generate electricity at about 30 kV. This is then converted to the grid voltage, usually 275 kV or 400 kV, using a step-up transformer.

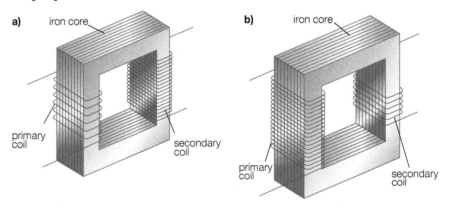

▲ **Figure 19.5** Transformers: **a)** step-up transformer **b)** step-down transformer

All transformers have three parts:

1 Primary coil – the input voltage is connected across this coil.
2 Secondary coil – the output voltage is across the secondary coil.
3 An iron core – this links the coils magnetically. No current flows in the iron core.

Summary

A step-up transformer:	A step-down transformer:
• increases the voltage	• decreases the voltage
• decreases the current	• increases the current
• has more turns on the secondary coil than on the primary.	• has more turns on the primary coil than on the secondary.

Between the power station and the grid, there is a step-up transformer to increase the voltage from around 30 kV to around 275 kV or 400 kV.

Why use high voltages?

The high voltages used to transmit electrical power around a country are extremely dangerous. That is why the cables that carry the power are supported on tall pylons high above people, traffic and buildings. Sometimes the cables are buried underground, but this is much more expensive, and the cables must be safely insulated.

High voltages are used because it means that the current flowing in the cables is relatively low. When a current flows in a wire or cable, some of the energy it is carrying is lost because of the cable's resistance, and the cable gets hot. A small current produces less heat and therefore wastes less energy than a large current.

Near the electricity user, the voltage is stepped down in stages using a number of step-down transformers. Heavy industry often takes its electricity at 415V, while equipment in a person's home usually requires the voltage to be reduced to a much safer 230V.

Test yourself

1 A centre-zero ammeter is connected to a coil of wire, as shown in Figure 19.6.
   a) How could you use this apparatus to induce a current in the coil?
   b) Suggest what you could do to maximise the size of the induced current.

2 Figure 19.7 shows the main stages in the generation, transmission and distribution of electricity.

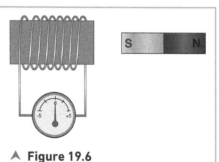

⋀ **Figure 19.6**

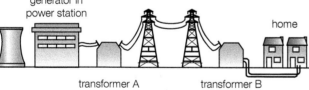

⋀ **Figure 19.7**

   a) In what way does transformer A alter the current?
   b) Explain briefly how the use of transformer A reduces the loss of energy in the overhead transmission lines.
   c) Describe fully the function of transformer B.

How long would it take you to name 20 electrical devices in your home? You could probably do that in less than a minute! The fact is that today, electricity is by far the most useful type of energy around. This is because it is the energy type that can be most easily changed into other energy forms.

Table 19.1 Common electrical **devices**

Device	Converts electricity into	Device	Converts electricity into
	light and heat		heat and light
	sound		heat
	kinetic energy		sound and light

An amusing story

The story is told (though no-one can be certain whether it is true) that after Faraday was made a fellow of the Royal Society, Mr. W.E. Gladstone, who would later become Prime Minister, asked what good his discovery of electromagnetic induction could be. "One day, sir" Faraday replied, "you may tax it!"

Renewable energy

Fossil fuel and nuclear energy are described as non-renewable because they will run out. There is a limited supply of each in the Earth's crust. That is why there has been emphasis in recent years on development of alternative renewable energy resources.

Renewable energy sources will not run out. The main sources are wind, waves, tides, solar and hydroelectric.

In wind turbines (Figure 19.8a), the blades act as the turbine. They turn the generator when the wind blows, producing electricity.

Wind energy is free, non-polluting and renewable. The disadvantages of using wind turbines are that they can be ugly and noisy, and they only produce electricity when there is a strong enough wind.

Solar cells (Figure 19.8b) directly convert light energy into electricity. Solar energy is free, non-polluting and renewable. The major disadvantage of solar cells is that they only produce electricity when they receive sunlight.

Solar panels can be placed on the roof of a building to provide hot water. Solar panels trap the heat energy from the Sun, as shown in Figure 19.8c.

Hydroelectric power involves building a dam to trap water (Figure 19.8d). Water is allowed to flow through tunnels in the dam. This turns turbines and thus drives generators. Once the dam is built, the energy is virtually free. No waste or pollution is produced. The major disadvantage is that the area behind the dam, which may formerly have provided habitats for endangered plants or animals, has to be flooded.

In Northern Ireland, short rotation coppicing of willow and poplar trees produces wood pellets for use in boilers to produce heat. These are renewable resources and carbon-neutral – the carbon dioxide used in photosynthesis is returned to the air on burning, so they take in as much carbon dioxide as they give off.

a)

b)

c)

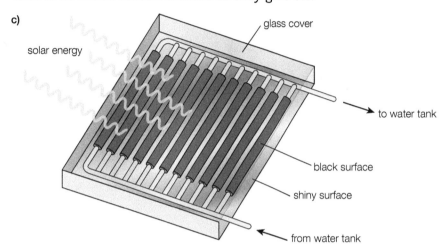

d)

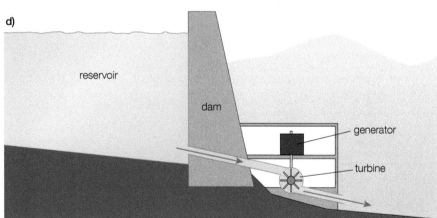

▲ **Figure 19.8 a)** A wind turbine **b)** A light powered by a solar cell **c)** A solar panel that can trap solar energy and use it to heat water **d)** A hydroelectric plant at a reservoir

Test yourself ✎

3 Name six devices that use electricity in the home, and state the useful output energy they produce.

Show you can ?

State the difference between renewable and non-renewable forms of energy and explain why there has been emphasis in recent years on the development of alternative renewable fuels.

1 An anemometer is a device to measure wind speed. Figure 19.9 shows a simple anemometer.

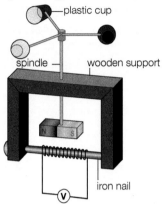

Figure 19.9

When the wind blows, the plastic cups turn.
a) Explain why the wind causes the voltmeter to give a reading. *(1 mark)*
b) Explain why the reading on the voltmeter is a 'measure' of the wind speed. *(1 mark)*
The gauge is not sensitive enough to measure the speed of a gentle breeze.
c) Suggest one way that the design can be modified to make the anemometer more sensitive. *(1 mark)*

2 A bar magnet is moving towards a loop of wire.

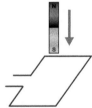

Figure 19.10

a) Is a voltage induced in the wire? Give a reason for your answer. *(2 marks)*
b) Is a current induced in the wire? Give a reason for your answer. *(2 marks)*

3 Figure 19.11 shows a bicycle dynamo.

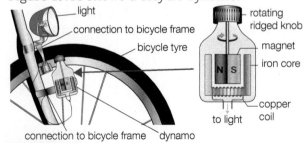

Figure 19.11

a) Explain fully why the lamp lights when the bicycle wheel turns. *(3 marks)*
b) Why does the lamp get brighter as the bicycle moves faster? *(3 marks)*
c) Why does the lamp not work at all when the bicycle stops? *(1 mark)*

4 Copy and complete the sentences below to show the main energy change which each device is designed to bring about.

An electric lamp converts electrical energy into .... energy. *(1 mark)*
A loudspeaker converts .... energy into .... energy. *(2 marks)*

5 Figure 19.12 shows a simple transformer.

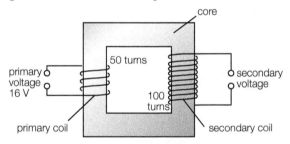

Figure 19.12

a) The primary and secondary coils are both wound on the same core. What is this core made of? *(1 mark)*
b) How can you tell from the diagram that this is a step-up transformer? *(1 mark)*

6 Figure 19.13 shows the main stages in the generation, transmission and distribution of electricity.

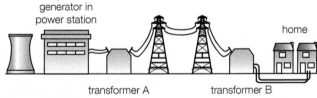

Figure 19.13

a) In what way does transformer A alter the voltage? *(1 mark)*
b) Explain why this voltage change is necessary. *(2 marks)*
c) In what way does transformer B alter the current? *(1 mark)*

20 Heat transfer

Heat transfer

There are three main methods of heat transfer.

1 Conduction – occurs mainly in solids. Most liquids are very poor conductors of heat and almost no heat conduction takes place in gases. For this reason, trapped air in fibreglass wool is an excellent insulator.

2 Convection – transfers heat only in liquids and gases.

3 Radiation – the only method of heat transfer in a vacuum.

Prescribed practical

Prescribed practical P3: Comparing the heat conductivity of different materials by measuring the time it takes to travel through a variety of conductors and at least one insulator

Aims
- to compare the conduction of heat in rods made of four different materials

Variables
- The independent variable is the material from which each rod is made.
- The controlled variables (which must be the same for each rod) are the length and cross-sectional area, the mass of wax used with each rod and the position of each rod in the flame.
- The dependent variable is the time taken for the pins to fall off each rod.

Apparatus
- rods of copper, aluminium, steel and glass
- a small candle
- 4 drawing pins
- a tripod on which to support the rods
- a Bunsen burner

Method
1 Prepare a suitable table for experimental results, as shown in Table 20.1.
2 Using a small amount of wax from the candle attach identical drawing pins to the end of each rod.

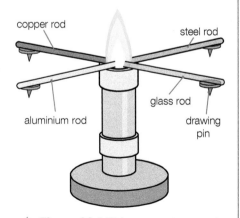

▲ **Figure 20.1** This apparatus can be used to demonstrate heat conduction in different materials

3 Support all four rods on a tripod in such a way that the ends not having the drawing pins will be at the same point in the Bunsen flame.

4 Light a Bunsen burner and adjust it to give a blue, but not a roaring, flame.

5 Place the Bunsen burner so that each rod is in the flame.

6 Immediately start a stop watch and time how long it takes for each pin to fall off.

7 Repeat the experiment to ensure reliability.

Table for results

Table 20.1

	Material	Copper	Aluminium	Steel	Glass
First trial	time taken for pin to fall/s				
Second trial	time taken for pin to fall/s				
Third trial	time taken for pin to fall/s				

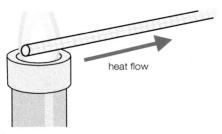

H ⌃ **Figure 20.2** In a metal rod, heat is conducted rapidly through the movement of free electrons

Heat conducts along all of the rods, but the pins fall off at different times. The pin attached to the copper rod falls first, shortly followed by the pin attached to the aluminium rod and the steel rod. Only after many minutes does the pin attached to the glass rod fall.

The experiment shows that copper and aluminium are good conductors but glass is a very poor heat conductor. Poor conductors are called insulators.

Conduction in metals

Free electron conduction

Most metals are good conductors of heat. Why is copper such a good conductor of heat? Unlike glass, copper has free electrons in its metallic structure. These are electrons that have escaped from atoms and can move freely throughout the solid. The free electrons absorb heat from the Bunsen flame (Figure 20.2). This heat allows them to move much faster than before.

As they move through the metal, free electrons collide with copper atoms. In these collisions, the electrons give some of their kinetic energy to the atoms and cause them to vibrate with greater amplitude than before. Free electron conduction is much faster than conduction caused by passing vibrations from atom to atom, so materials with free electrons (including all metals) are the best conductors of heat energy.

Conduction in insulators such as glass

In solids, the atoms are held together by chemical bonds. Although they cannot move around freely within the solid, they are able to vibrate. The part of the glass that is in the flame (Figure 20.3) absorbs heat energy. This makes the atoms in the end of the rod vibrate faster and with greater amplitude than their neighbours. These vibrations pass from atom to atom through the solid structure, transferring heat in the form of kinetic energy as they do so. Only after a considerable time does the energy of the flame reach the other end of the glass rod.

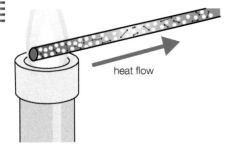

⌃ **Figure 20.3** In a glass rod, heat is conducted slowly as the vibrations pass from one atom to the next

Convection

Conduction in solids is always due to the movement of particles within the solid. The solid material itself does not move. However, when heat is transferred by convection in liquids and gases, the liquid and gas molecules themselves move to a different position.

Convection occurs when the fastest-moving particles in a hot region of a gas or liquid move to a cool region, taking their heat energy with them. It occurs only in liquids and gases because the atoms in solids are not free to move from place to place. Convection is explained by changes in the density of the material.

Convection in liquids

The movement of the purple dye in the water (Figure 20.4) shows the convection current.

As the water at the bottom of the flask warms up, the molecules gain kinetic energy. This extra energy causes the following to happen:

▶ The molecules vibrate with greater amplitude.
▶ This causes the warm water to expand.
▶ The density of the warm water is less than that of the cold water.
▶ The warm water rises.
▶ The cooler water flows downwards to replace the upward-moving warmer water.
▶ The cool water at the top falls as it is replaced by warm water.

Convection in air

Convection in the air can be demonstrated by the glass chimney experiment shown in Figure 20.5. First, a straw is lit and the flame is blown out. Note that the smoke rises when the straw is held in the air. Then the smoky straw is held over each chimney in turn.

When the smoky straw is over the candle flame, the smoke rises. When the straw is held over the other chimney, the smoke falls. If the straw is held in position for long enough, the smoke will eventually be seen in the horizontal section, and then rising above the candle flame.

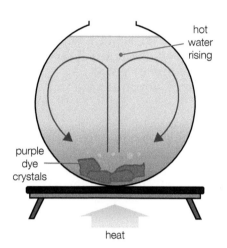

▲ **Figure 20.4** Demonstrating convection currents in a liquid

- purple dye crystals
- hot water rising
- heat

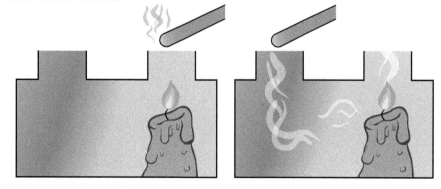

▲ **Figure 20.5** Demonstrating convection currents in air

Why does the smoke fall down the chimney?

▶ The air around the candle flame becomes very hot.
▶ The air molecules near the flame are moving faster than those in normal air.

Test yourself

1 Describe an experiment to show that metals are better conductors of heat than glass.
2 State the precautions you would take in the conduction experiment to ensure the test is fair.
3 Describe experiments that demonstrate convection in liquids and gases.

Show you can

1 Explain heat conduction in metals in terms of free electrons.
2 Explain heat conduction and convection in terms of the movement of particles.

▶ The hot air molecules are further apart than those in normal air because the air has expanded.

▶ The density of the hot air is less than that of normal air, so the hot air rises up the chimney.

▶ Cooler air moves along the horizontal tunnel to replace the air which has gone up the chimney.

▶ The moving smoke follows the motion of the cooler air.

Radiation

Radiation is the method of heat transfer that takes place without the need for any particles. It is the way by which the Earth receives heat energy from the Sun through the vacuum of space. The heat energy is transferred as infrared waves, which is part of the electromagnetic spectrum.

All objects radiate energy (emit radiant heat). The hotter an object is, the more radiation it emits. All objects also absorb radiant heat. If an object is hotter than its surroundings, it emits more radiant heat than it absorbs, so its temperature falls. If an object is cooler than its surroundings, it absorbs more radiant heat than it emits, so its temperature rises.

Giving out radiation (radiation emission)

Figure 20.6 shows an experiment in which a thick piece of copper is covered with gloss (shiny) white paint on one side and matt (non-shiny) black paint on the other. The copper has been heated with a Bunsen burner until it is very hot.

If you were to hold your hand about 30 cm from the gloss white side, and then hold your hand about the same distance from the black side, you would notice that your hand would feel much hotter facing the matt black surface. This is because the matt black surface is the better emitter of radiant heat.

Absorption of radiation

Figure 20.7 shows two sheets of thin aluminium, one painted gloss white and the other matt black. A cork is fixed to the back of each vertical plate with candle wax as 'glue'. The plates are placed equal distances away from a Bunsen burner. When the burner is lit, each plate receives the same quantity of radiant heat. The wax on the matt black plate will melt first, and the cork will fall off. The white plate warms much more slowly, so the cork takes much longer to fall off. This is because the black surface is the better absorber and the white surface is the better reflector of radiant heat.

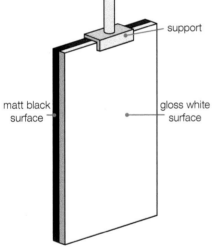

▲ **Figure 20.6** Investigating radiation emission

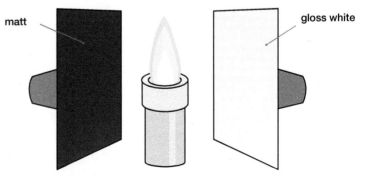

▲ **Figure 20.7** Investigating radiation absorption

Radiation summary

▶ Black surfaces are the best emitters and best absorbers of radiation.

▶ White surfaces are the worst emitters and worst absorbers of radiation.

▶ Matt surfaces are better emitters and better absorbers of radiation than gloss surfaces.

Air – nature's insulator

What happens when we hold a match a few centimetres away from a flame as shown in Figure 20.8? The heat reaching the match head is not enough to light the match. This is because air is a very poor conductor of heat.

The heat arrives at the match head almost entirely by radiation. Hot air rises, so almost no heat arrives by convection. Air is a very poor conductor of heat, so no heat arrives by conduction.

If we held the match the same distance vertically above the Bunsen flame, it would burst into flame almost immediately. This is because the match would be receiving heat by convection as well as by radiation.

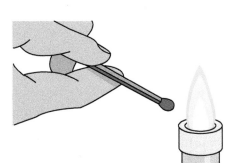

▲ **Figure 20.8** Why won't the match light?

Applications of heat transfer

Vacuum flasks

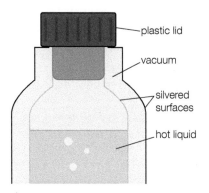

▲ **Figure 20.9** A cross-section through a vacuum flask

The vacuum flask shown in Figure 20.9 was designed by James Dewar in order to keep liquids cold. But the flask works equally well as a way of keeping liquids hot. Today, it is commonly used as a picnic flask to keep tea, coffee or soup hot. How does it work?

▶ The flask is made of a double-walled glass bottle. There is a vacuum between the two walls. The vacuum stops all heat transfer by conduction or convection through the sides.

▶ The glass walls facing the vacuum are silvered. Their shiny surfaces reduce heat transfer by radiation to a minimum.

▶ The stopper is made of plastic, and is often filled with cork or foam to reduce heat transfer by conduction through it.

▶ The outer, sponge-packed plastic case protects the inner, fragile flask against physical damage.

Solar panels

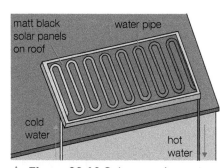

▲ **Figure 20.10** Solar panels

A solar panel absorbs sunlight and uses the energy to heat water (Figure 20.10) in the following process:

▶ The sunlight passes through a glass window and falls onto a blackened metal sheet.

▶ The metal is in a draught-proof enclosure to minimise heat loss by convection.

▶ The blackened metal absorbs almost all of the energy in the sunlight and its temperature often rises to over 100 °C.

▶ The heat stored in the blackened metal is then transferred to water flowing in a nearby pipe.

▶ Solar panels are ideal for use where large volumes of hot water are needed, such as swimming pools and hospital laundries.

Reducing heat loss from your home

Heat from a home is lost through the roof, walls, windows, floor and so on.

Different materials and devices have been designed to reduce this heat loss. Table 20.2 gives some of the methods used to reduce heating bills in our home. The pay back time is how long it takes for the savings on heating bills to equal the cost of installation. The pay back time for a series of common installations is also shown in Table 20.2.

Table 20.2 Reducing heat loss from our homes

Device	Pay back time	How losses are reduced
cavity wall insulation	3 years	the cavity between the outside walls is filled with fibreglass, mineral wool or foam. these materials trap air in tiny pockets. trapped air reduces heat loss through walls by convection and conduction.
loft (attic) insulation	1.5 years	fibreglass or mineral wool fibres trap air. trapped air reduces heat loss through the roof by convection and conduction.
double glazing	40 years	two panes of glass are placed in a frame with a small gap between them. the trapped air reduces heat loss by convection and conduction.
thick curtains and carpets	variable (depends on quality)	reduce draughts. trapped air reduces heat loss through windows and floors by convection and conduction.
jacket on hot water tank	1 year	the jacket is made of an insulating material such as fibreglass or mineral wool. mineral wool and foam both trap air in tiny pockets. this trapped air reduces heat loss through walls by convection and conduction.
draught-proofing	1 year	reduces flow of warm air, minimising heat loss by convection.
thermostatic controls on radiators	5 years	keeps the room at the required temperature, reducing waste of fuel due to over-heating.

Show you can

1 Describe an experiment to show that dark matt surfaces are better emitters of radiant heat energy than light shiny surfaces.
2 Describe an experiment to show that dark matt surfaces are better absorbers of radiant heat energy than light shiny surfaces.
3 Describe and explain two everyday applications of heat transfer.
4 Describe three ways by which heat energy can be lost from our homes and how this heat loss can be reduced.

Test yourself

4 The cost of thermostatic controls on all the radiators in a house is £150. How much money in reduced fuel bills would this save in a year? (Hint: Look at the pay back period in Table 20.2.)
5 A double glazing salesman claims that his product will save the householder £60 each year in reduced fuel bills. Calculate the cost of this double glazing and comment on whether it would be the best investment to reduce fuel bills.
6 Draught-proofing a house saves a householder £65 each year in reduced fuel bills. What is the cost of the draught-proofing?
7 Which of the methods of reducing heat loss in the home would you recommend to a young family on a limited budget? Give reasons for your answers.

Practice questions

1 A pot of stew is being heated on a gas cooker.
   a) What is the main method of heat transfer through the metal base of the pot? *(1 mark)*
   b) What is the main method of heat transfer through the stew? *(1 mark)*
   c) What particle is responsible for heat transfer through the wooden handle of the saucepan? *(1 mark)*
   d) What method of heat transfer is reduced by making the outside surface of the copper pot shiny? *(1 mark)*

2 A vacuum flask is constructed to keep liquids hot. Copy and complete Table 20.3 by adding ticks to show which methods of heat transfer are reduced by each of the features labelled in Figure 20.11. *(3 marks)*

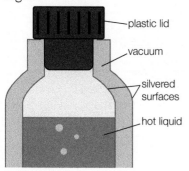

Figure 20.11

Table 20.3

Feature	Conduction	Convection	Radiation
plastic lid			
silvered surfaces			
vacuum			

3 a) Copy and complete Figure 20.12 by adding the required labels. *(2 marks)*

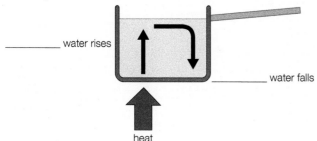

Figure 20.12

   b) What name is given to the movement of the water as it is heated in this way? *(1 mark)*
   c) Explain why this movement takes place. *(3 marks)*

4 An iron rod and a glass rod are placed in a Bunsen flame as shown in Figure 20.13. The dimensions of both rods are exactly the same.

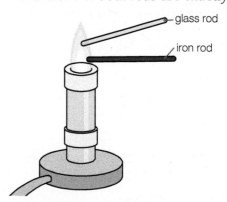

Figure 20.13

Copy Table 20.4. After each statement, write the letter G if the statement applies to glass, write I if the statement applies iron, and write GI if the statement applies to both rods.

Table 20.4

This rod has no free electrons.	
Atoms are mainly responsible for heat conduction.	
Atoms vibrate more quickly when heat is added.	
Heat is transferred when electrons collide with neighbouring atoms.	

(4 marks)

5 Two conical flasks contain the same amount of water at the same temperature. They are placed at equal distances from an electrical heater.

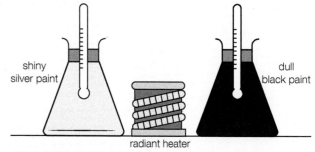

Figure 20.14

   a) By what process is heat mainly transferred from the heater to the conical flasks? *(1 mark)*
   b) In which flask does the temperature rise more quickly? Explain fully the reasons for your answer. *(2 marks)*

6 Container ships are used to carry fruit and vegetables all over the world. The food must be kept cool to prevent it from spoiling. The hold of a container ship has two metal walls with an insulator between them.
 a) What is the purpose of the insulating material? *(1 mark)*
 b) Name a suitable insulator for this purpose. *(1 mark)*
 c) What makes this insulator effective? *(1 mark)*

7 a) Which part of an oven is hotter, the top or the bottom? Give a reason for your answer. *(2 marks)*
 b) What is the purpose of the fan in a fan-assisted oven? *(1 mark)*

8 Computer chips can produce a lot of heat. If they become too hot, they might get damaged.

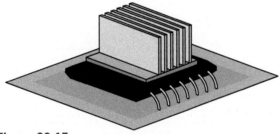

Figure 20.15

They are often attached to a heavy piece of copper metal with black fins, as shown in Figure 20.15. Such a piece of metal is called a heat-sink. What four features of the design of this type of heat-sink make it suitable for its purpose? *(8 marks)*

9 Figure 20.16 shows a solar panel.

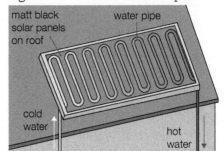

matt black solar panels on roof

water pipe

cold water

hot water

Figure 20.16

The sunlight passes through a glass window and falls onto a blackened metal sheet.
 a) Suggest why it is necessary to have a window. *(1 mark)*
 b) Why has the metal sheet been blackened? *(1 mark)*

The heat stored in the blackened metal is then transferred to water flowing in a nearby copper pipe.
 c) Why is the pipe made of copper? *(1 mark)*
 d) By what process is the heat transferred into the copper? *(1 mark)*

Solar panels are ideal for use where large volumes of hot water are needed.
 e) Suggest what type of institution might be interested in installing solar panels. *(1 mark)*

21 Waves

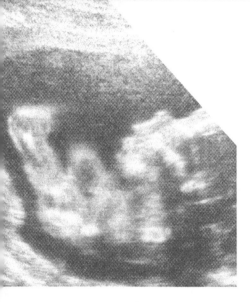

Types of waves

Waves transfer energy from one point to another, but they do not, in general, transfer matter. Radio waves, for example, carry energy from a radio transmitter to your home, but no matter moves in the air as a result.

All waves are produced as a result of vibrations, and they can be classified as longitudinal or transverse. A vibration is a repeated movement, first in one direction and then in the opposite direction.

Longitudinal waves

A longitudinal wave is one in which the particles vibrate parallel to the direction in which the wave is travelling. The only types of longitudinal waves relevant to your GCSE course are:

▶ sound waves
▶ ultrasound waves
▶ slinky spring waves.

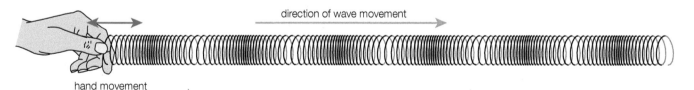

direction of wave movement

hand movement

▲ **Figure 21.1** A longitudinal wave moving along a slinky spring

It is easy to demonstrate longitudinal waves by holding a slinky spring at one end and moving your hand backwards and forwards parallel to the axis of the stretched spring as shown in Figure 21.1. Compressions are places where the coils (or particles) bunch together. Rarefactions are places where the coils (or particles) are furthest apart.

All longitudinal waves are made up of compressions and rarefactions. In the case of sound waves, the particles are the molecules of the material through which the sound is travelling. These molecules bunch together and separate just as they do in a longitudinal wave on a slinky spring.

Transverse waves

A transverse wave is one in which the vibrations are at 90° to the direction in which the wave is travelling. Most waves in nature are transverse – some examples are:

▶ water waves
▶ slinky spring waves
▶ electromagnetic waves.

A transverse wave pulse can be created by shaking one end of a slinky spring (Figure 21.2). The pulse moves along the slinky, but the final position of the slinky is exactly the same as it was at the beginning. None of the material of the slinky has moved permanently. But the wave pulse has carried energy from one point to another.

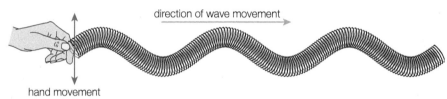
direction of wave movement

hand movement

▲ **Figure 21.2** Transverse waves in a slinky spring

You can see that water waves are transverse. A cork floating on the surface of some water bobs up and down as the waves pass (Figure 21.3). The vertical vibration of the cork is perpendicular to the horizontal motion of the wave. Energy is transferred in the directions in which the wave is travelling.

There are many other examples that show that waves carry energy:

▶ visible light, infrared radiation and microwaves all make things heat up
▶ X-rays and gamma waves can damage cells by disrupting DNA
▶ loud sound waves can cause objects to vibrate (even if that is only your eardrum)
▶ water waves can be used to generate electricity.

> **Tip** ↻
>
> Slinky springs can be used to demonstrate both longitudinal and transverse waves, so it is best not to use these as an example of a longitudinal (or a transverse) wave.

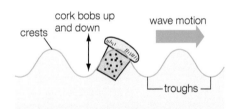
crests | cork bobs up and down | wave motion
troughs

▲ **Figure 21.3** Water waves transfer energy to cork

Show you can

1 Describe the difference between a transverse wave and a longitudinal wave.
2 Give two examples of transverse waves and two examples of longitudinal waves.
3 Describe the evidence to show that microwaves transmit energy.
4 Describe how you could use a slinky spring to demonstrate:
    a) a longitudinal wave
    b) a transverse wave.

Wave features

Wavelength

The wavelength (λ) is the distance between two successive crests or troughs, as shown in Figure 21.4. λ is the Greek letter l (for wavelength) and is pronounced 'lamda'. Wavelength is the length of one complete vibration and is measured in millimetres (mm), centimetres (cm) or metres (m).

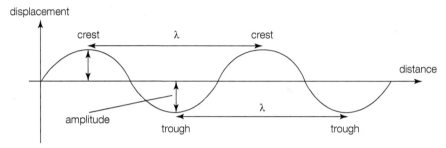

▲ **Figure 21.4** Displacement–distance graph to illustrate wavelength and amplitude

Amplitude

The amplitude (A) is the maximum height of the wave. Amplitude is measured from the mid-line to the crest (or trough) in millimetres (mm), centimetres (cm) or metres (m).

Frequency

The frequency (f) is the number of vibrations that pass a particular point in one second. Frequency is measured in hertz (Hz).
1 vibration per second = 1 Hz.

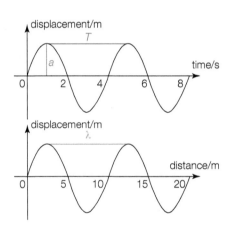

▲ **Figure 21.5** Graphs of waves

Graphs and waves

We can represent waves on graphs like in Figure 21.5.

Notice that the upper graph is displacement against time. The lower graph is displacement against distance. The vertical axis on both graphs is displacement, so we can find the amplitude from either graph.

The red line in the upper graph shows the time, T, between the crests passing a fixed point. This time is known as the period. Students often wrongly think T is a distance (the wavelength).

The blue line in the lower graph shows the distance between consecutive crests. This is the wavelength, λ.

The period, T, in the upper graph is 4 seconds, and the wavelength, λ, in the lower graph is 10 metres.

How could we find the speed from these data?

The graphs tell us that the wave is travelling 10 metres in 4 seconds.

We know that speed = $\dfrac{\text{distance}}{\text{time}}$

So, the wave speed $= \dfrac{10\,\text{m}}{4\,\text{s}} = 2.5\,\text{m/s}$

We can also find the frequency from the top graph. The period (the time taken for one wave to pass), T, is 4 seconds, so 0.25 waves must pass every second.

So the frequency, $f = \dfrac{1}{T} = 0.25\,\text{Hz}$

We can use the frequency to confirm the speed using the wave equation:

$v = f \times \lambda$

$v = 0.25\,\text{Hz} \times 10\,\text{m}$

$\quad = 2.5\,\text{m/s}$

The wave equation

Imagine a wave with wavelength λ (metres) and frequency f (hertz).

The speed of the wave, v, is given by:

wave speed = frequency × wavelength

$v = f \times \lambda$

Note that the units used in this equation must be consistent, as shown in Table 21.1.

Table 21.1 Units for frequency, wavelength and speed

Frequency	Wavelength	Speed
always in Hz	cm	cm/s
	m	m/s
	km	km/s

Test yourself

1 Copy and complete Table 21.2. Note carefully the unit in which you are to give your answers. The first one has been done for you as an example.

Table 21.2

Wavelength	Frequency	Speed
5 m	200 Hz	1000 m/s
12 m	50 Hz	.... m/s
3 cm	60 kHz	.... m/s
...... m	4 Hz	20 cm/s
...... m	5 kHz	2.5 km/s
16 mm	...... Hz	80 cm/s
6×10^4 m	...... Hz	3×10^8 m/s

Show you can

1 The vertical distance between a crest and a trough is 24 cm, and the horizontal distance between the first and the fifth wave crests is 40 cm. If 30 such waves pass a fixed point every minute, find the amplitude, frequency, wavelength and speed of the waves.

1 Find the speed of a sound wave of frequency 1.1 kHz and wavelength 30 cm.

Answer

f = 1.1 kHz

= 1100 Hz

λ = 0.3 m

$v = f \times \lambda$

= 1100 Hz × 0.3 m

= 330 m/s

2 Find the frequency of radio waves of wavelength 1500 m if their speed is 300 000 km/s.

Answer

v = 300 000 km/s

= 300 000 000 m/s

$v = f \times \lambda$

so

300 000 000 = f × 1500

$$f = \frac{300\,000\,000}{1500}$$

= 200 000 Hz

= 200 kHz

Sound

▶ Sounds are produced by vibrating objects.
▶ Sound is a longitudinal wave.
▶ Sound can travel through solids, liquids and gases, but cannot travel through a vacuum (empty space).
▶ Sound travels most quickly through solids and most slowly through gases.
▶ Sound travels at 330 metres per second through air.
▶ Bigger vibrations make louder sounds. This can be seen on a graph as a bigger amplitude (waves A and B, Figure 21.6).
▶ Slower vibrations produce sounds of lower frequency or pitch. The lower the frequency, the longer the wavelength (waves C and D, Figure 21.6).

Hearing

The frequency range of human hearing is from 20 Hz to 20 kHz. This is called the audible hearing range. As we get older our upper limit decreases. Older people can find it difficult to hear sounds above 14 000 Hz (14 kHz).

A
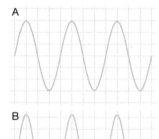
B
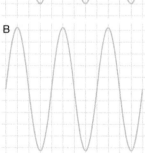
C
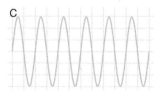
D

▲ **Figure 21.6** Wave A is quieter than wave B as its amplitude is smaller. Wave D has a lower frequency than wave C, so its wavelength is longer.

The factors that affect a person's hearing include:
- birth defects
- damage to the eardrum
- prolonged exposure to a noisy environment
- age.

Earplugs and protectors are required in noisy environments and on building sites to prevent hearing damage.

Echoes

Like all waves, sound can be reflected off a surface. A reflected sound is called an echo. The best reflectors of sound are hard surfaces. This is why the bathroom and kitchen are usually the 'loudest' rooms in your home. Soft surfaces are good absorbers of sound.

Echoes can be both very useful and a real nuisance. They are a nuisance in concert halls and theatres, where the music reaches the audience not only from the stage but also as echoes created by the walls, floor and ceiling. This causes reverberation and can reduce quality. The problem is overcome in auditoria by using soft, sound-absorbing materials in the walls and ceilings.

Ultrasound

Sounds with a frequency higher than 20 000 Hz (20 kHz) are known as ultrasound. The human ear cannot hear these sounds.

Uses of ultrasound

Ultrasound is used extensively in medicine because it is much safer than X-rays.

Ultrasound is used in medicine for:
- scanning the developing foetus in the womb (Figure 21.7)
- early diagnosis (e.g. to detect the presence of gall stones in the gall bladder)
- checking for internal bleeding in an organ
- scanning veins and arteries for clots
- removing plaque from teeth in dentistry
- observing the working of the heart in real time (e.g. the opening and closing of cardiac valves).

Ultrasound is used extensively in industry to:
- clean delicate articles of jewellery, watches and lenses
- create the controllable heat needed to weld plastics
- detect micro-cracks in railway lines
- measure distances (e.g. the depth of the ocean or the distance from one end of a room to the other)
- detect fish and submarines.

Bats use echoes and ultrasound to navigate and to find the flying insects that they eat. Dolphins use ultrasound to find their prey.

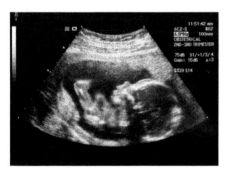

▲ **Figure 21.7** Ultrasound is used to scan the foetus in the womb

Example

1 A pulse of ultrasound takes 4 seconds to travel from a dolphin to its prey and back again. The speed of sound in sea water is 1500 m/s.

Calculate how far away the dolphin is from its prey.

Answer

distance = speed × time

      = 1500 m/s × 4 s

      = 6000 m

This is the distance from the dolphin to its prey and back again, so the distance between the dolphin and its prey is 3000 m.

2 A ship sends out a pulse of ultrasound that takes 0.6 seconds to return.

The speed of sound in sea water is 1500 m/s.

How deep is the seabed beneath the surface?

Answer

distance = speed × time

      = 1500 m/s × 0.6 m

      = 900 m

This is the distance from the ship to the seabed and back again, so the seabed is 450 m beneath the surface.

Measuring the speed of sound

Two methods are described – the echo method and the flash-bang method.

The echo method

Two students go outside to estimate the speed of sound. One hits two wooden blocks together.

The other student measures the time between the bang and hearing the echo (Figure 21.8). The wall is 300 m away and the average time found was 1.8 seconds.

▲ **Figure 21.8** Using the echo method to measure the speed of sound

The time to the wall and back = 1.8 s, so the time to the wall = 0.9 s.

$$\text{speed} = \frac{\text{distance}}{\text{time}}$$
$$= \frac{300\,\text{m}}{0.9\,\text{s}}$$
$$= 333\,\text{m/s}$$

The flash-bang method

H

For this experiment, one person stands at a point in a large field. A second person drives a car a set distance, at least one kilometre away and turns it to face the first person. The person in the car then simultaneously flashes the car's headlights and sounds the horn. When the lights are seen, the person standing starts a stop watch (Figure 21.9). They stop it when the sound of the horn is heard.

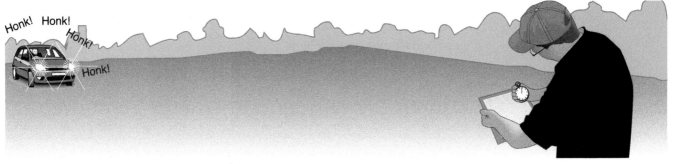

Honk! Honk! Honk! Honk!

▲ **Figure 21.9** The flash-bang method to find the speed of sound

This is repeated several times, and an average is found of the time for the sound to arrive. This reduces the effect of human error and increases the experiment's reliability. To reduce possible error due to the wind, the two people reverse positions and measure the time of travel of the sound in the opposite direction. The speed of sound is then taken as the average of the two calculated speeds.

Typical results are shown in Table 21.3.

Table 21.3 Measuring the speed of sound by the flash-bang method

	Distance travelled by sound/m	Average time taken/s	Speed/ m/s	Average speed/ m/s
with the wind	990	2.9	341	330
against the wind	990	3.1	319	

Test yourself

2 State what causes sound.
3 State the audible range of human hearing and describe how it changes as we get older.
4 Describe the difference between sound and ultrasound.
5 Describe what is meant by an echo.
6 State three different uses of ultrasound.

Electromagnetic waves

Electromagnetic waves are members of a family with common properties called the electromagnetic spectrum. They:

▶ can travel in a vacuum (unique property of electromagnetic waves)

▶ travel at exactly the same speed in a vacuum

▶ are transverse waves.

Electromagnetic waves also show properties common to all types of wave. They:

▶ carry energy

▶ can be reflected

▶ can be refracted.

There are seven members of the electromagnetic family. The properties of electromagnetic waves depend very much on their wavelength. In Table 21.4, they are arranged in order of increasing wavelength (or decreasing frequency) along with their dangers and uses.

The electromagnetic waves that can cause cancer are gamma rays, X-rays and ultraviolet light. These are the electromagnetic waves that have the highest frequency and carry the most energy. It is because they have so much energy that these particular waves are capable of disrupting the DNA in living cells and causing cancer.

Dangers and uses of electromagnetic waves

Table 21.4 shows the electromagnetic spectrum, and lists the dangers and uses of each type of wave.

Table 21.4 The electromagnetic spectrum

Electromagnetic wave	Dangers	Uses
Gamma (γ) rays	can kill cells; can disrupt DNA, which may lead to cancer	• cancer treatment • sterilisation of equipment
X-rays	can kill cells; can disrupt DNA, which may lead to cancer	• food preservation • cancer treatment • diagnosis of broken bones and dental problems
Ultraviolet light	can cause skin cancer	• sun beds • detection of forged banknotes
Visible light	intense visible light can damage the eyes	• allows us to see • photosynthesis • helps skin cells produce vitamin D • used in telephone networks in optical fibres
Infrared light	can cause burns	• heaters • security (PIR) detectors • remote controls • optical fibre communications
Microwaves	some scientists think that microwaves can cause internal heating of body tissues which could potentially lead to eye cataracts	• cooking food • fast satellite communications • mobile phones • speed cameras
Radio waves	are thought to potentially cause cancer when experienced in very large doses	• long range communication • radio and TV broadcasting

How do microwaves heat up food?

Most foods contain some water. Microwaves have continuously changing electric and magnetic fields associated with them, and can readily penetrate food. When a microwave comes across a water molecule, the electric field causes the water molecule to oscillate about two billion times a second. This rapid movement of the water molecule is observed as an increase in temperature. Microwaves therefore work best when heating foods that contain a lot of water. This is why they work so well when re-heating ready meals, soups and stews in microwave ovens.

▲ **Figure 21.10** A microwave oven

Mobile phones

A mobile phone sends and receives messages carried by microwaves. The signal is sent from the mobile phone to the nearest mast. From there, it is passed on to the next mast and so on until it reaches its destination (Figure 21.11).

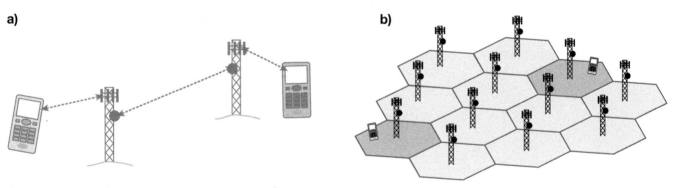

a)

b)

▲ **Figure 21.11 a)** How a mobile phone works **b)** A cell network

The area around a mobile phone mast is called a cell. That is why in the USA mobile phones are called cell phones. Each cell acts as a repeater station. Without these repeater stations, the signals would have to be much stronger, which would make mobile phones much bigger. Long-distance calls will involve satellite transceivers (Figure 21.12).

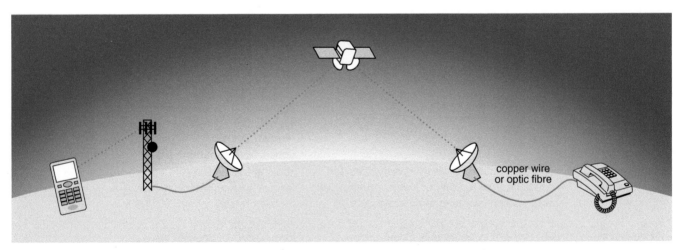

copper wire
or optic fibre

▲ **Figure 21.12** Long-distance calls involving satellite transceivers

Health risks of mobile phones and communication masts

Mobile phones and mobile phone masts emit and receive microwaves, and many people have suggested there may be a link between microwave radiation and cancer. There are claims that because mobile phones use microwaves, holding the phone close to your ear can cause the brain to be damaged.

Independent expert groups set up by Britain, the USA and the European Union have reported that there is no proven case of damage being done to people either by communication masts or by mobile phones. However, most expert groups have recommended the precautionary principle is followed. This states that even if the chances of negative health effects are low, it makes sense to avoid unnecessary risk.

This means:

▶ Young children should use mobile phones very sparingly, because their small body mass would make any possible harm to them more severe.

▶ People should be encouraged to use mobile phones with headsets or speaker-phones whenever possible, so as to keep their heads as far as possible from the radiation-emitting handset.

▶ Mobile phone masts should not be erected close to schools or hospitals.

Practice questions

1 A student uses a stretched slinky to make longitudinal waves.

Figure 21.13

a) What do the longitudinal waves transfer from A to B? *(1 mark)*
b) Copy the diagram and on it, draw a double-headed arrow to indicate the direction that the student would have to move end A to make longitudinal waves. *(1 mark)*
c) The student sends 18 waves along the slinky in 6 seconds. How many waves does she make per second? *(1 mark)*
d) Use your answer to part (c) to state the frequency of the waves. *(1 mark)*
e) What is the wavelength of these longitudinal waves? *(1 mark)*
f) Use your answers to parts (d) and (e) to calculate the speed of these longitudinal waves. *(3 marks)*
g) Give another example of a longitudinal wave. *(1 mark)*

2 Figure 21.14 shows a wave.

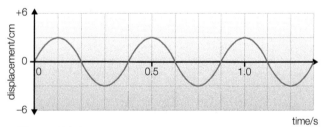

Figure 21.14

a) Copy the graph, and on it draw another wave with the same frequency but twice the amplitude of the one shown. *(2 marks)*
b) Explain what is meant by ultrasound. *(2 marks)*
c) Figure 21.15 shows a car using ultrasound to help it park.

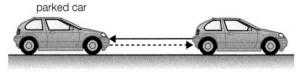

Figure 21.15

The car sends an ultrasound signal to the parked car.
This signal is reflected, and is received 0.01 s after being sent.
The speed of ultrasound in air is 330 m/s.
Calculate the distance between the cars. *(3 marks)*

3 a) Explain fully how the rays in a microwave oven heat food. *(4 marks)*
b) Microwaves travel in air at a speed of 3×10^8 m/s. Calculate the wavelength of microwaves of frequency 1250 MHz. Give your answer in centimetres. *(3 marks)*

4 In an experiment on hearing, sounds of different frequencies were played to 20 teenagers and 20 pensioners. The number who could hear each frequency was recorded. The results are shown in Table 21.5.

Table 21.5

Frequency/kHz	12	14	16	18	20	22
number of teenagers who could hear this frequency	20	20	20	20	20	0
number of pensioners who could hear this frequency	20	18	15	12	0	0

a) Describe fully what the information in the table tells us about hearing in:
   i) teenagers
   ii) pensioners. *(4 marks)*
b) What name is given to sounds above 20 kHz? *(1 mark)*

5 A car and a group of students are 3 km apart. A person in the car flashes the headlights and sounds the horn at the same time. When the students see the light from the car, they immediately start their stopwatches. They stop their stopwatches when they hear the car's horn.
a) What information does this give us about the speed of sound compared to the speed of light? *(1 mark)*
b) Suggest a reason why the students do not all obtain the same value in their time measurement. *(1 mark)*
The average time shown on the stopwatches is 9.1 seconds.
c) Calculate the speed of sound from this data. *(3 marks)*

The car and the students then change positions and the experiment is repeated.
The value obtained for the speed of sound is now 350 m/s.

d) Suggest a reason why the speed of sound is different in one direction than in the other.

(1 mark)

6 Figure 21.16 shows some members of the electromagnetic spectrum arranged in order of increasing wavelength.

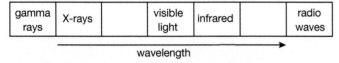

gamma rays	X-rays		visible light	infrared		radio waves

wavelength

Figure 21.16

a) Which two members of the electromagnetic spectrum are missing? *(2 marks)*

b) State a property unique to electromagnetic waves. *(1 mark)*

c) State one use and one danger of:
  i) X-rays
  ii) infrared radiation. *(4 marks)*

22 Force and motion

Fossil fuels for transport

The most important fuels used for transport today are petrol and diesel. These fuels come from crude oil, which is a fossil fuel. Less frequently used is LNG (liquefied natural gas) which is also a fossil fuel. These fuels are all non-renewable, which means that one day we are going to run out of them. It is important that we search for alternative renewable fuels and find ways of extending the life of existing reserves of fossil fuels. A breakdown of the fuels used in the UK from 2011 to 2015 can be seen in Table 22.1.

Renewable fuels

Biofuel

Biofuel can be produced from oil extracted from the seeds of oilseed rape. It is almost carbon neutral, contributing less than fossil fuels to the greenhouse effect. The plants from which biofuels are made take in as much carbon dioxide as the fuel gives out when it is burned, which is why they are largely 'carbon neutral'.

Biodiesel

Biodiesel is a renewable diesel oil made from a mix of oils and fats including recycled cooking oil, soybean oil, and animal fats. It can be used as it is in existing diesel engines without modification. It is often added to ordinary diesel oil rather than used alone. In the UK, the use of biodiesel has fluctuated considerably over recent years.

Gasohol

In Brazil and Thailand, the oil companies produce gasohol by adding bioethanol (alcohol) to petrol. This extends their oil supplies. Bioethanol is made from the fermentation of sugar beet, rice, wheat, barley, sugar cane, potatoes, corn and other biomass materials. In the UK, the use of bioethanol has increased irregularly since 2011, with peak production in 2014.

Other new fuels include hydrogen gas and methanol.

Table 22.1 Use of fuels in road transport in the UK (numbers in thousands of tonnes)

Year	2011	2012	2013	2014	2015
Petrol	13895	13231	12574	12326	12082
Bioethanol	517	615	650	645	631
Diesel	20991	21538	21926	22675	23656
Biodiesel	825	563	682	850	595
Gas	92	94	94	88	92

Greener transport

Regenerative hybrid systems

For many years, London buses used regenerative braking systems. The buses had electrical generators attached to the wheels. When the bus braked, its kinetic energy was converted to electricity. This recharged the battery.

The Toyota Prius and several other cars have an ordinary petrol engine and a high-voltage rechargeable battery. As far as possible, all wasted energy is used to recharge the battery, which can be used at times to power the vehicle. The Nissan Leaf is a car that has no fossil fuel engine at all. It runs on electricity from rechargeable batteries.

Using fuel cells

Ordinary batteries convert chemical energy into electrical energy. When the chemicals inside the battery are used up, the battery is returned for recycling. Fuel cells are a special type of battery (Figure 22.1) in which the chemicals are supplied continuously by the user, in much the same way as petrol is supplied to keep a car engine working.

A car using a fuel cell is an electric car in which the user tops up the battery by adding the fuel as needed. Many car manufacturers are currently researching the use of fuel cells to power their vehicles.

Fuel cells can use many different chemicals. One type uses methanol, which is readily made from ordinary alcohol and steam. The alcohol needed to make methanol is produced from renewable biomass. The use of renewable methanol fuel can replace fossil fuels directly. Inside the cell, the methanol 'burns' to produce electricity. The waste products are carbon dioxide, water and heat. Liquid methanol fuel can be purchased from some filling stations in the same way as petrol.

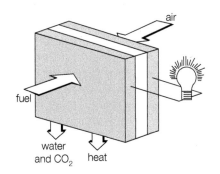

▲ **Figure 22.1** How a fuel cell works

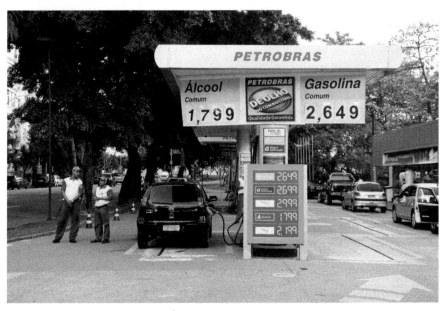

▲ **Figure 22.2** Buying methanol fuel at a filling station

Show you can

1 Describe some attempts being made to minimise reliance on fossil fuels for transport.
2 Explain why biodiesel is (almost) carbon neutral.
3 Explain what regenerative braking means.
4 In the context of road transport, explain what hybrid systems are.

Road safety

One cause of road accidents is driving too close to the vehicle in front. If the leading vehicle brakes suddenly, there could be a crash. According to the Highway Code, you should leave enough space between you and the vehicle in front to pull up safely if it suddenly slows down or stops (Figure 22.3). The safe rule is never to get closer than the overall stopping distance.

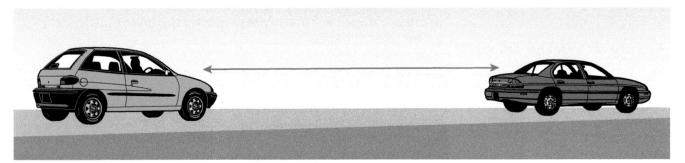

▲ **Figure 22.3** The stopping distance should always be less than the distance to the car in front

Stopping distance

The stopping distance is how far a vehicle travels from the time the driver sees a reason to stop until the vehicle reaches a complete stop. It is the sum of the thinking distance and the braking distance.

Thinking distance

The thinking distance is how far the vehicle travels while the driver is thinking about what to do. It is the distance the car travels from when the driver sees the danger until the brakes are applied.

Braking distance

The braking distance is how far the vehicle travels after the brakes are applied until it comes to a complete stop.

Table 22.2 gives the typical thinking, braking and stopping distances at different speeds, on dry roads with good tyres and brakes.

Table 22.2 Typical thinking, braking and stopping distances

Speed/mph and m/s	Thinking distance/metres	Braking distance/ metres	Stopping distance/metres
20 mph 9 m/s	6	6	12
40 mph 18 m/s	12	24	36
60 mph 27 m/s	18	55	73

Factors affecting stopping distances

What affects thinking distance?

Anything that slows the driver's brain will increase the thinking distance. It will take the driver longer to think what they should do, and the car will travel further in this time. The major factors that increase thinking distance are:

▶ speed – the higher the speed, the greater the distance the vehicle travels in a certain time
▶ alcohol – slows down the brain and impairs judgement
▶ tiredness – slows down the brain and impairs judgement
▶ illegal drugs – slow down the brain and impair judgement
▶ prescribed medicines – may slow down the brain. If there is any danger, you will be warned by the chemist's label.

What affects braking distance?

Anything that reduces the amount of friction or the grip of the brakes and between the tyres and the road will increase the braking distance. The major factors that increase braking distance are:

▶ speed – the higher the speed, the greater the distance the vehicle travels in a certain time
▶ state of tyres – bald tyres give less grip between the tyres and the road
▶ weather – wet or icy roads give less grip between the tyres and the road
▶ state of the brakes – poor brakes reduce the grip between the brakes and the wheels
▶ state of the road – poor road surfaces reduce grip between the tyres and the road.

summer tyre

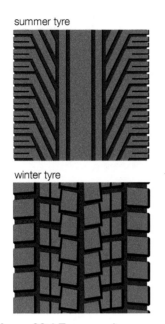

winter tyre

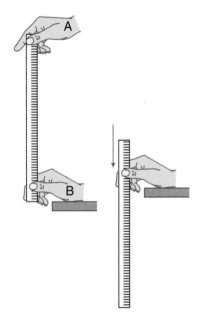

▲ **Figure 22.4** Tyres can have different treads for different conditions

It is important to be aware that the data for braking and stopping distance in Table 22.2 are for vehicles with good tyres and brakes travelling on dry roads.

Friction

Friction is the force that opposes motion. It is measured in newtons (N). You should learn this definition.

Braking distance depends on friction between the tyres and the road. That is why the condition of the tyres on a vehicle is so important. Different treads can be used to optimise grip at different times of year (Figure 22.4).

Friction is produced when two surfaces rub together. The rougher the surfaces, the more friction is produced. Friction depends only on the nature of the surfaces in contact and the weight of the object.

Reaction time

When reading about stopping distance, you learned that it has two parts – thinking and braking distances. Thinking distance is how far the vehicle travels while the driver is thinking what to do. It depends on the speed of the driver's brain.

Reaction time is the time that passes between an observation and the start of the body's response to that observation. Reaction times can be measured roughly using a metre ruler, as shown in Figure 22.5.

The metre ruler is held vertically by person A against person B's outstretched hand. The zero on the metre ruler, must be level with the tip of B's index finger. Person A lets go of the metre ruler, and person B tries to catch it as quickly as possible. The distance the ruler falls can be used to calculate reaction time using the following equation:

$$t^2 = \frac{d}{4.9}$$

where d = distance travelled by the metre ruler in metres and t = reaction time in seconds.

To ensure a reliable, accurate result, this experiment must be carried out several times and an average of all the results should be taken.

Show you can ❓

1 Explain what is meant by **a)** thinking distance **b)** braking distance and **c)** stopping distance.
2 List three factors which affect **a)** thinking distance and **b)** braking distance.
3 Explain what is meant by friction.
4 Describe an experiment to measure a person's reaction time.

Test yourself

1 What is meant by a fuel extender?
2 How is bioethanol made?
3 Why is it important for a motorist to drive more than their stopping distance behind the car in front?
4 A motorist is travelling at 20 m/s. If his reaction time is 0.9 s, calculate his thinking distance.

▲ **Figure 22.5** Using a metre ruler to measure reaction time

Physics and car accidents

Crumple zones are areas at the front and rear of a car that are designed to absorb the huge amount of kinetic energy lost in an accident (Figure 22.6). Crumple zones collapse relatively easily and slowly. They spread the collision over a longer period of time, and so reduce the force on the passengers. This helps to reduce the likelihood of serious injuries occurring.

By absorbing the energy of the impact, crumple zones reduce the risk of serious injuries to the people in the vehicle.

The car's cabin is much stronger. Around the driver and passengers, there is a rigid passenger cell that is designed not to crumple in the event of an accident. This reduces the risk of serious injury to the people inside the car.

Seat belts provide a restraining force and reduce the risk that drivers and passengers will be thrown forward and seriously injured. Drivers and passengers must, by law, wear seatbelts (Figure 22.7).

▲ **Figure 22.6** Crumple zones make forces on car passengers smaller

▲ **Figure 22.7** Drivers and passengers, by law, must wear seatbelts

▲ **Figure 22.8** Airbags are fitted in most modern cars

Front airbags are fitted in the steering wheel or in the dashboard (Figure 22.8). The shock of a frontend collision sets off a controlled, explosive chemical reaction inside the bag. The reaction generates a large volume of gas very quickly. The gas fills the bag, which then holds a passenger in their seat. The bags are porous and go down quickly after the accident. Many cars also have airbags fitted to the sides. These reduce the number of crush injuries that occur from the side impact of another vehicle.

Motorway **crash barriers** are designed to prevent vehicles crossing from one carriageway to the other. They also prevent vehicles from impacting or entering roadside hazards. The barriers absorb some of the kinetic energy from the impact caused by the vehicle striking them. They also redirect the vehicle along the line of the barrier so that it does not turn around, turn over or re-enter the stream of traffic. Motorway designers call this Road Traffic Accident Containment.

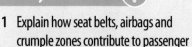

Show you can

1 Explain how seat belts, airbags and crumple zones contribute to passenger safety in the event of a car accident.

2 Explain what a rigid passenger cell is and describe how it contributes to passenger safety.

Speed limits

The speed limit on any road depends on the type of road and type of vehicle. The speed limits for UK roads are summarised in Table 22.3. The two main reasons for speed limits are:

▶ to make accidents less likely
▶ to reduce the risk of fatal injury if an accident occurs.

Table 22.3 Speed limits for roads in the UK

Type of vehicle	In built-up areas/ mph	On single carriageways/mph	On dual carriageways/mph	On motorways/ mph
cars and motorcyles	30	60	70	70
cars towing trailers or caravans	30	50	60	60
buses and coaches	30	50	70	70
goods vehicles	30	50	70	70
heavy goods vehicles (HGVs)	30	40	60	60

Police and local councils attempt to enforce these speed restrictions, using a variety of methods.

Speed cameras

Speed cameras are used to photograph speeding vehicles and those jumping traffic lights. There are hand-held (Figure 22.9), instantaneous speed (Figure 22.10), and average speed varieties. An instantaneous speed camera measures the time taken to travel a very short distance.

▲ **Figure 22.9** A hand-held speed camera

▲ **Figure 22.10** An instantaneous speed camera

Instantaneous speed is shown on a car's speedometer. It shows the speed of the car at that moment in time. The owner of a speeding vehicle will be identified by means of the registration number. The driver may have to pay a fine and get penalty points on their licence.

Speed bumps

Speed bumps, as shown in Figure 22.11, are used to slow vehicles down in built-up areas. Drivers are forced either to slow down or risk a bumpy ride with possible damage to their vehicle.

Calming measures

Speed cameras, speed bumps and road-narrowing schemes are commonly called traffic calming measures. The advantage of these measures is that drivers are forced to slow down and this leads to fewer and less serious accidents. The major disadvantage of such measures is that they increase the time taken to provide emergency services.

Calculating average speed

Hand-held cameras always calculate your instantaneous speed. However, some speeding motorists slow down just before the camera and so avoid detection. To prevent this, many police forces use average speed cameras. These cameras work in pairs. The two cameras are a known distance apart. A clock starts when the driver passes one camera, and stops when they pass the second camera. The difference gives the time taken to travel the known distance. From these two measurements, the average speed can be found using the equation:

$$\text{average speed} = \frac{\text{total distance travelled}}{\text{time taken}}$$

▲ **Figure 22.11** A speed bump used to slow vehicles down in built-up areas

▲ **Figure 22.12** A sign warning motorists that average speed cameras are being used

---

Example

On a road where the speed limit is 30 miles per hour, the time taken for a motorist to travel the 1800 metres between two speed cameras is 125.0 s.

a) Calculate the motorist's average speed in metres per second.

b) If 1 m/s equals 2.24 mph, use your answer to part a) to calculate the motorist's speed in miles per hour and state whether or not he was breaking the speed limit.

Answer

a) $\text{average speed} = \dfrac{\text{total distance travelled}}{\text{time taken}}$

$= \dfrac{1800}{125}$

$= 14.4 \text{ m/s}$

b) $14.4 \text{ m/s} = 14.4 \times 2.24 \text{ mph}$

$= 32.3 \text{ mph}$

The motorist was over the 30 mph speed limit.

Distance–time graphs

Physicists often like to show data graphically. Figure 22.13 shows how the distance travelled by a pedestrian changes with time. There are four rules to help you interpret such graphs.

▶ Horizontal lines mean the object is stationary (not moving, zero speed).

▶ Diagonal lines mean the object is moving with constant speed.

▶ The gradient (slope) of a straight line is equal to the constant speed.

▶ The steeper the diagonal line, the faster the object is moving.

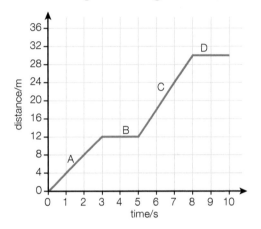

▲ **Figure 22.13** Distance–time graph for a pedestrian

Example

Using the graph in Figure 22.13:

a) Identify the region where the pedestrian was travelling fastest.

b) Calculate the speed of the pedestrian in regions A and C.

c) Identify the regions where the pedestrian was stationary.

d) Calculate the average speed for the entire journey.

Answer

a) The pedestrian was travelling fastest in region C because that is where the line is steepest.

b) In A, the pedestrian travels 12 m in 3 s, so the speed is 12 m ÷ 3 s = 4 m/s.

In C, the pedestrian travels 18 m (from 12 m to 30 m on the vertical axis) in a time of 3 s (from 5 s to 8 s on the horizontal axis), so the speed is 18 m ÷ 3 s = 6 m/s.

c) The pedestrian was stationary in regions B and D because the lines here are horizontal.

d) average speed = $\dfrac{\text{total distance travelled}}{\text{time taken}}$

$= \dfrac{30}{10}$

$= 3\,\text{m/s}$

Test yourself

5 a) Figure 22.14 shows the distance–time graph for a cyclist.

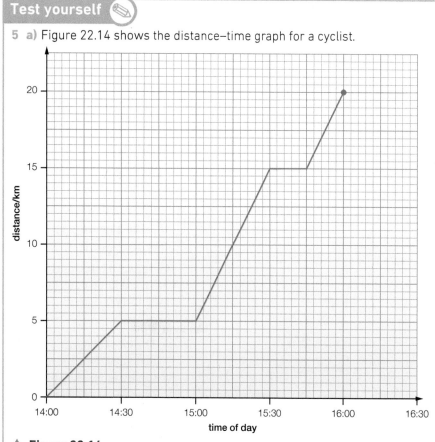

▲ **Figure 22.14**

    i) Describe the motion of the cyclist between 14.30 and 15.00.

    ii) Between which times is the cyclist travelling at the fastest speed?

    iii) At what speed would the cyclist have to travel to cover the same total distance in 2 hours?

b) Another cyclist starts to travel at 20 km/hour at 14.45. This cyclist starts from the same point as the first.

    i) Copy the graph. On the same graph, plot how the distance changes with time for the second cyclist.

    ii) At what time do the two cyclists meet?

Balanced and unbalanced forces

If the forces on an object are equal in size but opposite in direction, the forces are said to be balanced.

If the forces on an object are balanced, its velocity remains constant. This is known as Newton's first law of motion.

This means that if the resultant force on an object is zero, the object will either remain at rest or move in a straight line with a constant speed.

The unbalanced force on an object is known as the resultant force. It causes the object's speed or direction to change. The object therefore accelerates. This is illustrated in the next two examples.

1 A lorry is moving at a constant speed in a straight line.

backward force   forward force

a) The forward force due to the engine is 30 000 N.

   i)   Which one of the following statements is correct?
        - The backward force is less than 30 000 N.
        - The backward force is equal to 30 000 N.
        - The backward force is greater than 30 000 N.

   ii) What is the name of the backward force acting on the lorry?

b) The forward force due to the engine increases to 50 000 N. If the backward force does not change, what happens to the speed of the lorry?

Answer

a) i)   The lorry is moving at a constant speed, so the forces are balanced. This means the backward force and the forward force are equal in size. The backward force is 30 000 N.

   ii) Friction

b) The forces are now unbalanced so the lorry accelerates, getting faster and faster.

2 A car is travelling in a straight line on a level road. The force of friction at the wheels is 2000 N. The force provided by the car's engine is 2400 N.

a) Calculate the resultant (unbalanced) force, if any, on the car.

b) Describe the motion of the car.

The car then passes over a stretch of rough road, so that the friction force increases from 2000 N to 2400 N. The engine force is unchanged.

c) Calculate the new resultant (unbalanced) force, if any, on the car.

d) Describe the motion of the car.

Answer

a) Resultant (unbalanced) force = 2400 – 2000 = 400 N forwards

b) The car is accelerating as there is a forward resultant force of 400 N.

c) There is now no resultant (unbalanced) force. The engine force (2400 N) and friction force (2400 N) are the same in size, but opposite in direction. The forces are balanced.

d) The car moves at a steady speed in a straight line.

6 State the formula for average speed.

Acceleration

The acceleration is the rate of change of the velocity of an object. If an object is getting faster, it is accelerating. If an object is slowing down, it is decelerating. If the object is getting faster, we say its acceleration is positive. If it is slowing down, its acceleration is negative.

Acceleration is measured in m/s^2. An acceleration of 2 m/s^2 means that every second, the velocity increases by 2 m/s. An acceleration of -3 m/s^2 means that every second, the velocity decreases by 3 m/s.

Newton discovered around 1686 the relationship between the resultant force F on an object and its acceleration. This relationship is commonly called Newton's second law of motion. Expressed as an equation, Newton's second law of motion can be written:

$$F = m \times a$$

where F is the resultant force in N, m is the mass of the object in kg and a is the acceleration in m/s^2.

Example

A car has a mass of 800 kg. Its engine produces a forward force of 1520 N and the friction force is 320 N.

a) Calculate:

  i) the size of the resultant force

  ii) the car's acceleration.

b) The driver applies the brakes and the car decelerates at 3 m/s^2. Calculate the resultant force.

Answer

a) i) resultant force = forward force − friction force

$$= 1520 - 320$$
$$= 1200 \text{ N}$$

  ii) $a = \dfrac{F}{m}$

$$= \frac{1200}{800}$$
$$= 1.5 \text{ m/s}^2$$

b) acceleration = -3 m/s^2 (Note the acceleration is negative because the car is slowing down.)

$$F = ma$$
$$= 800 \text{ kg} \times -3 \text{ m/s}^2$$
$$= -2400 \text{ N}$$

Note that the force is negative because it is opposing the direction of the motion.

1 Figure 22.15 shows the total stopping distances for cars travelling at different speeds.

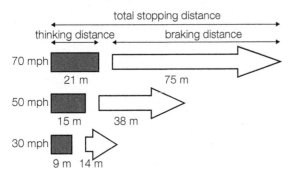

Figure 22.15

a) Calculate the total stopping distance at 70 mph. *(1 mark)*

b) Using the diagram, state two effects of increasing speed. *(2 marks)*

c) Copy and complete Table 22.4 to show what effect (if any) each condition may have on thinking and braking distance. Choose from these words:

increased   none   decreased *(2 marks)*

Table 22.4

Condition	Thinking distance	Braking distance
wet road surface		
worn tyres	none	increased
driver has been drinking alcohol		

H 2 Copy and complete the following statements by adding the appropriate words or phrases.
Petrol and diesel are examples of _____ .
Biofuel can be produced from _____ extracted from the seeds of oilseed rape.
Biodiesel is a renewable diesel oil made from a mix of oils and fats including _____ and _____ .

Bioethanol is made from the fermentation of materials like _____ and _____
Bioethanol is an example of a fossil fuel _____ , because it increases the lifetime of existing stocks of fossil fuels.
Other new fuels used for road transport are _____ and _____ . *(9 marks)*

3 Figure 22.16 shows Julie's journey to school.

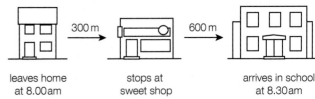

leaves home at 8.00am   stops at sweet shop   arrives in school at 8.30am

Figure 22.16

a) What distance (in metres) did Julie walk to school? *(1 mark)*

b) How long, in minutes, did it take Julie to get to school? *(1 mark)*

c) Use the equation to calculate Julie's average speed on her journey from home to school. *(2 marks)*

$$\text{average speed} = \frac{\text{total distance travelled}}{\text{time taken}}$$

d) If Julie does not stop at the shop but walks straight to school, what, if anything, will happen to her average speed? *(1 mark)*

4 Robin is investigating friction. He uses a newtonmeter to measure the force needed to pull a block of wood across different surfaces.

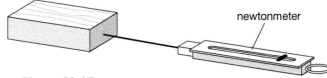

newtonmeter

Figure 22.17

Here are his results.

Table 22.5

Surface	Force needed/N
wooden bench	2.5
tiled floor	2
tarmac	4
carpet	3.5

a) Which surface has the lowest friction? *(1 mark)*

b) What could Robin do to reduce the friction of the wooden bench? *(1 mark)*

c) Robin gives the block the same size of push across each surface. On which surface would the block travel the shortest distance? Explain your answer. *(1 mark)*

H 5 During a race, a cyclist of mass 60 kg accelerated at 1.5 m/s² on a bicycle of mass 10 kg. The friction force is 15 N.

a) i) Use the equation: resultant force = mass × acceleration to calculate the resultant force which acts on the cyclist and his bicycle.
(3 marks)

ii) Calculate the forward force produced by the cyclist.
(2 marks)

b) A bullet of mass 25 g is fired into a piece of wood in a forensic testing laboratory. The average resistive force acting on the bullet as it becomes embedded in the wood is 196 N. Show that the deceleration of the bullet is 7840 m/s².
(3 marks)

6 A gymnast pulls himself up a rope.

rope

Figure 22.18

a) At the moment shown in Figure 22.18, he is stationary. What is the name of the upward force exerted on the gymnast? *(1 mark)*

b) The gymnast now slides down the rope at a constant velocity. What can you now say about the upward force? Choose your answer by placing a tick in the correct box. *(1 mark)*

Table 22.6

The upward force is now zero.	
The upward force is now equal to the gymnast's weight.	
The upward force is greater than the gymnast's weight.	

227

23 Radioactivity

Structure of the atom and radioactivity

All matter is made up of tiny particles called atoms. Inside every atom, there are three types of particle; protons, neutrons and electrons (Figure 23.1). The properties of these particles are summarised in Table 23.1.

Table 23.1 Properties of particles within the atom

Particle	Location	Relative mass*	Relative charge*
proton	within the nucleus	1	+1
neutron	within the nucleus	1	0
electron	orbiting the nucleus	$\dfrac{1}{1840}$	−1

*Mass and charge are measured relative to the proton

Radioactive elements have atoms with large, unstable nuclei. They are unstable because they contain too many protons or too many neutrons. These unstable nuclei will disintegrate or decay (split up) into smaller, more stable nuclei. When this happens, the nuclei emit radiation. There are three types of radiation, which we call alpha (α), beta (β) and gamma (γ).

nucleus

- e (electron)
- P (proton)
- N (neutron)

shells

▲ **Figure 23.1** The structure of an atom

Background radiation

Radioactive material is found naturally all around us, and even inside our bodies. This type of radiation is called background radiation.

Sources of background radiation include:

▶ carbon-14, which is a type of carbon found inside all living organisms

▶ a radioactive gas called radon, which is emitted from granite rocks

▶ measurable traces of potassium which are contained in some foods

▶ waste products from nuclear power stations

▶ radioactive materials used in medicine in hospitals

▶ cosmic rays striking the Earth from space.

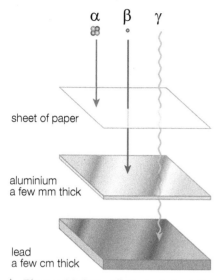

sheet of paper

aluminium
a few mm thick

lead
a few cm thick

⋀ **Figure 23.2** Protecting against the three different types of radiation

Types of radiation

The three types of radiation that can be emitted from radioactive nuclei are summarised in Table 23.2 and Figure 23.2.

Table 23.2 Types of radiation

Name	Description	Stopped by	Relative charge
alpha (α)	a positively charged helium nucleus – a heavy, slow particle	a few centimetres of air or a thin sheet of paper	+2
beta (β)	a small, light and fast moving electron	several metres of air or a thin sheet of aluminium	−1
gamma (γ)	electromagnetic radiation of very high energy	easily passes through paper and aluminum, but can be blocked by lead	0

Half-life

Radioactivity is a completely random process. We cannot say which atom in a piece of radioactive material will disintegrate next, but we can count the number of disintegrations per second.

The term half-life is used to describe how radioactive a substance is. The shorter the half-life, the faster the material is decaying, and so the more radioactive it is.

The half-life of a radioactive material is the time taken for its activity to fall to half of its original value. Learn this definition so that you can quote it accurately in an exam.

This means that in one half-life period, the number of radioactive atoms will decrease by half (Figure 23.3).

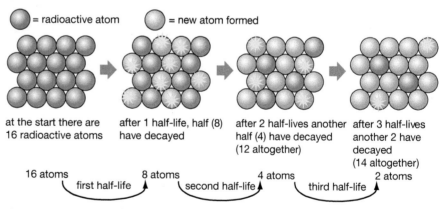

⬤ = radioactive atom   ⬤ = new atom formed

at the start there are 16 radioactive atoms

after 1 half-life, half (8) have decayed

after 2 half-lives another half (4) have decayed (12 altogether)

after 3 half-lives another 2 have decayed (14 altogether)

16 atoms — first half-life → 8 atoms — second half-life → 4 atoms — third half-life → 2 atoms

⋀ **Figure 23.3** Explaining the meaning of half-life

The greater the half-life of a radioactive source, the more slowly it decays, and so the longer it will take to become safe. Some sources have half-lives of tens of thousands of years, so it will take a very long time indeed for these materials to become safe. But what exactly do we mean by 'become safe'?

There is no agreed answer as to what is meant by 'safe'. But, in general, a radioactive material is not considered safe if the activity around it is above background radiation. In Northern Ireland, the background radiation level is about 15 disintegrations per minute.

Unit of radioactivity

The activity of a radioactive material is often measured in units called becquerels (Bq). One disintegration per second is 1 becquerel. So if a radioactive material emits 1200 α-particles every minute, then 20 nuclei decay every second, and its activity is 20 Bq.

Taking background into account

Suppose a physicist is trying to measure the radioactivity coming from cherries in an orchard. His instrument detects not only the radiation coming from the cherries, but it also detects the background radiation. To measure the activity of the cherries themselves, the physicist must subtract the background count from the measured count. So if the instrument showed a total activity of 43 Bq and the background count was 15 Bq, the activity due to the cherries alone would be 43 − 15 = 28 Bq.

▲ **Figure 23.4** When measuring the activity of a material, remember to subtract the background count

---

Example

1   The half-life of cobalt-60 is approximately five years. A particular source has an activity of 4000 Bq.

   a) How long will it take for this source's activity to decay to 500 Bq?

   b) What would be the activity of the source after 25 years?

Answer
This kind of problem is best solved with a table.

Activity/Bq	Time/half-lives	Time/years
4000	0	0 (now)
2000	1	5
1000	2	10
500	3	15
250	4	20
125	5	25

Notice that in the table the activity halves every half-life (that is, every 5 years).

a) It takes 15 years for the activity to fall to 500 Bq.

b) The activity after 25 years is 125 Bq.

2 Nitrogen-13 is radioactive. A physicist has a sample of 160 g of nitrogen-13.

After 40 minutes, only 10 grams of the sample remains. The remainder has decayed.
Calculate the half-life of nitrogen-13.

Answer
Again, we use a table. The mass of nitrogen-13 remaining halves in every half-life.

Mass/g	Time/half-lives	Time/minutes
160	0	0 (now)
80	1	
40	2	
20	3	
10	4	40

In 40 minutes, it takes 4 half-lives to decay to 10 g, so the half-life is 40 ÷ 4 = 10 minutes.

3 When a radioisotope arrives at a hospital, its activity level before correcting for the background count is 1030 Bq. Calculate the half-life of the source if the activity before correcting for background has fallen to 48 Bq in 60 hours. The background count is 16 Bq.

Activity/Bq	Corrected activity/Bq	Time/half-lives	Time/hours
1030	1024	0	0 (now)
	512	1	
	256	2	
	128	3	
	64	4	
48	32	5	60

Notice that to obtain the corrected activity, we subtract the background count (16 Bq).

Only the corrected activity is halved every half-life.

Since the table shows it takes five half-lives for the uncorrected activity to fall to 48 Bq, the half-life is 60 ÷ 5 = 12 hours.

Show you can

1 Calculate what fraction of a radioactive source remains undecayed after three half-lives.
2 Demonstrate how background activity is taken into account in calculations.

Test yourself

1 Explain the cause of radioactivity and state the unit in which it is measured.
2 State the nature and origin of alpha particles, beta particles and gamma radiation.
3 State the penetrating powers of alpha particles, beta particles and gamma radiation.
4 Explain what is meant by background radiation and state some of its sources.
5 Define the term half-life.

H

Ionising radiation

When alpha (α), beta (β) or gamma (γ) radiation or X-rays collide with atoms, they are most likely to hit the outer electrons, knocking them out of their orbit.

All atoms are electrically neutral as they have equal numbers of protons and electrons. This means that there are equal numbers of positive and negative charges. When electrons are knocked out, the atom becomes positively charged as it loses negative charge. This is illustrated in Figure 23.5.

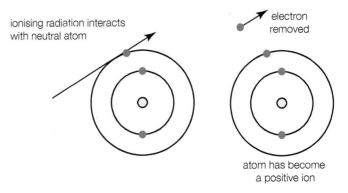

ionising radiation interacts with neutral atom

electron removed

atom has become a positive ion

▲ **Figure 23.5** Showing what is meant by ionisation

Charged atoms are called ions, and the process of turning neutral atoms into charged ions is called **ionisation**. This is why alpha, beta and gamma radiation are also called **ionising radiation**.

When radiation ionises the molecules of living cells, it can cause cancers and kill healthy cells. The larger the dose of radiation, the greater the risk. Outside the body, beta and gamma radiation are more dangerous than alpha radiation. Inside the body, an alpha source causes the most damage because it is the most ionising. This is summarised in Figure 23.6.

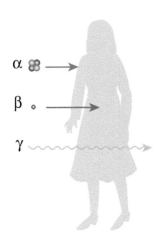

α

β

γ

▲ **Figure 23.6** Radiation and the human body

Minimising the risk from ionising radiation

There are, in principle, three ways to reduce the risk from ionising radiation.

▶ Maximise the distance between the person and the source.

▶ Minimise the exposure time.

▶ Shield the observer from the source.

Maximising the distance from the source is why radioactive sources are always held with tongs (or forceps) at arm's length from the user.

Minimising exposure time is the reason why radioactive sources are always kept in locked cupboards well away from the laboratory, and only taken out when actually being used. Even then, it is good practice to work quickly with radioactive sources, to minimise the user's exposure to radiation.

Shielding is particularly appropriate when using very powerful sources of radiation. Figure 23.7 shows a typical radiotherapy treatment room such as might be found in a large hospital. The windows and walls are all lined with lead to absorb any unwanted radiation. The therapists remain behind these shields when the machine is switched on.

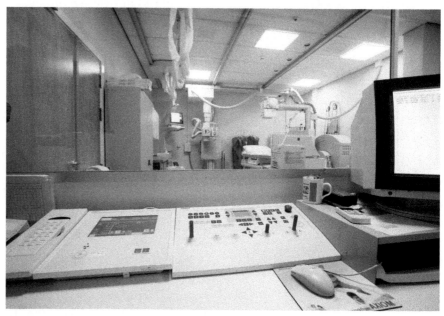

▲ **Figure 23.7** Shielding against ionising radiation in a hospital

Uses of radioactivity in industry, medicine and agriculture

Industry

In industry, radiation is used to control the thickness of metals such as aluminium as it passes through a rolling mill, as shown in Figure 23.8. If the count from the detector is too small, the sheet metal is too thick, so the signal is sent to the rollers to increase the pressure. If the count from the detector is too large, the sheet metal is too thin, so the signal is sent to the rollers to decrease the pressure.

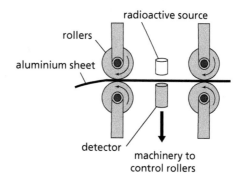

▲ **Figure 23.8** Controlling metal thickness in aluminium

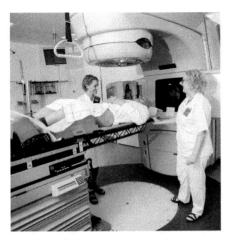

▲ **Figure 23.9** A radiotherapist at work

Medicine

Radiotherapy is the use of radiation in medicine to kill cancer cells, as shown in Figure 23.9. Gamma rays are often used because they can penetrate deep inside the body. These powerful gamma rays are targeted at the tumour, as shown in Figure 23.10. The radiation source is usually rotated to have maximum effect on the cancer cells, but not destroy the healthy cells surrounding the tumour.

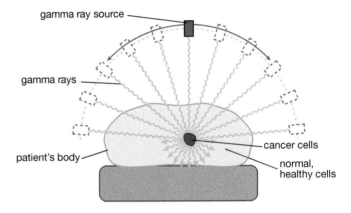

▲ **Figure 23.10** Rotating the gamma ray source

Gamma radiation is also used in medicine and dentistry to sterilise surgical instruments. It will kill any bacteria, viruses or fungi found on them.

Agriculture

In the food and agriculture industry, gamma radiation can also be used to treat fresh food. The radiation will kill any bacteria and fungi on the food, stopping it from decaying.

This means the food will stay fresher for longer, extending its shelf-life.

▲ **Figure 23.11** Gamma radiated strawberries on the left and untreated strawberries on the right

Practice questions

1 **a) i)** What is meant by radioactivity? *(1 mark)*
   ii) Explain fully why some nuclei are radioactive. *(1 mark)*
   b) i) What is meant by background radiation? *(1 mark)*
   ii) Give two sources of background radiation. *(2 marks)*

Figure 23.12 shows how the activity of a radioactive isotope varies with time.

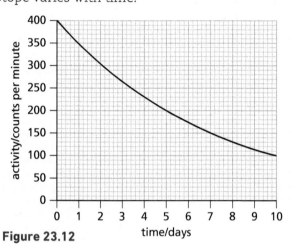

Figure 23.12

   c) i) What is the activity at 7 days? *(1 mark)*
   ii) Describe the trend shown by this graph. *(1 mark)*

2 Use the graph in question 1 to find the half-life of the radioactive source. *(1 mark)*

3 Copy and complete Table 23.3 to show the properties of alpha (α), beta (β) and gamma (γ) radiation.

Table 23.3

Radiation	Comes from	Nature of radiation	Stopped by
α	unstable nucleus		
β		fast moving electron	
γ			thick sheet of lead

(6 marks)

4 A radioisotope arrives in a hospital. Six days later, it is found that its activity is 4096 Bq. After a further 24 days, its activity has fallen to 256 Bq.
   a) Calculate the half-life of this radioisotope. *(3 marks)*
   b) Show that on the day of arrival in the hospital its activity was 8192 Bq. *(1 mark)*

5 Outline one use of radioactivity in:
   a) industry
   b) medicine
   c) agriculture. *(3 marks)*

6 Three types of radiation, alpha, beta and gamma, are emitted from radioactive sources.
   Copy and complete Table 23.4 by writing the name of the relevant radiation in the space provided. *(4 marks)*

Table 23.4

Property	Radiation
can be stopped by thin paper	
emitted from a nucleus and has a negative charge	
is an electromagnetic wave	
is a positively charged helium nucleus	

7 **a)** What is meant by the word ionisation? *(1 mark)*
   b) Alpha radiation is much more ionising than beta or gamma radiation. Suggest why sources of alpha radiation outside the human body are least likely to cause damage to major organs inside the body. *(2 marks)*
   c) State three precautions that might be taken by people who work with ionising radiation to minimise the risk to themselves. *(3 marks)*

8 Radioactive materials can be used to check for leaks in water pipes as shown in Figure 23.13. The radioisotope soaks into the ground at the point where the pipe leaks. The radiation count directly above the leak is high, but on either side of the leak the count is low.

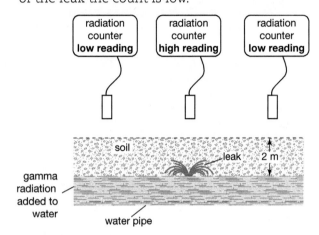

Figure 23.13

Six radioisotopes are available. Their properties are shown in Table 23.5.

Table 23.5

Radioisotope	Radiation emitted	Half life
A	alpha (α)	24 hours
B	alpha (α)	7 days
C	beta (β)	15 minutes
D	beta (β)	12 hours
E	gamma (γ)	15 seconds
F	gamma (γ)	3 hours

Which one of these radioisotopes (A – F) is **most** suitable for this purpose?
Give **two** reasons for your answer.   *(2 marks)*

24 The Earth in space

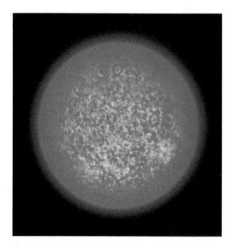

Specification points

This chapter covers 3.9.1 to 3.9.10 of the specification.
It includes the variety of objects that make up our Solar System and the forces that keep these heavenly bodies in orbit. You will learn about the processes of star formation, red shift and Big Bang theory.

The Solar System

Our Solar System, shown in Figure 24.1, is made up of one star called the Sun, and eight planets called Mercury, Venus, Earth, Mars, Jupiter, Saturn, Uranus, and Neptune. It also includes comets, asteroids, moons and smaller objects called meteoroids. The planets travel around the Sun in paths called orbits. All the planets travel around the Sun in the same direction and in the same plane.

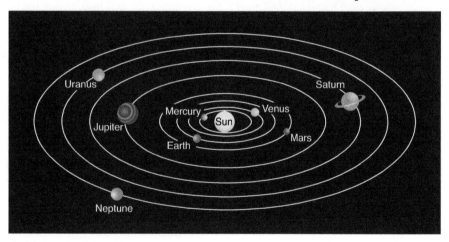

▲ **Figure 24.1** The Solar System

The four inner planets (Mercury, Venus, Earth and Mars) orbit the Sun in almost circular paths. The inner planets have a hard surface and are called rocky planets. It is possible to land a spacecraft on these inner planets.

The outer planets have elliptical orbits, shaped roughly like the side view of a rugby ball. These outer planets are much bigger than the inner planets and are very large, very dense balls of gas. They are sometimes called the gas giants.

Objects that are held in an orbit are called satellites. So the planets are all satellites of the Sun. Other satellites of the Sun are comets and asteroids. More information on the planets is given in Table 24.1.

▲ **Figure 24.2** The Sun

Table 24.1 Some data on the eight planets orbiting the Sun

Planet	Planet diameter compared with Earth	Average distance of planet from Sun compared with Earth	Time to orbit the Sun compared with the Earth	Number of moons
Mercury	0.4	0.4	0.2	0
Venus	0.9	0.7	0.6	0
Earth	1.0	1.0	1.0	1
Mars	0.5	1.5	1.9	2
Jupiter	11.2	5.2	12.0	14
Saturn	9.4	9.5	29.0	24
Uranus	4.1	19.1	84.0	15
Neptune	3.9	30.1	165.0	3

▲ **Figure 24.3** A comet

▲ **Figure 24.4** The Barringer Crater in the USA was probably caused by an asteroid strike 50 000 years ago. It is about 1,200 m in diameter and 170 m deep.

Comets are large chunks of frozen rock covered by huge quantities of frozen water and gases. They orbit the Sun, just like planets, but have orbits that are much more elliptical than that of a planet. They can only be seen when they pass close to the Sun. The Sun's energy causes some of the frozen gases and water to vaporise. This creates the comet's 'tail', as shown in Figure 24.3.

Asteroids are large chunks of rock that orbit the Sun. Most are found in the asteroid belt between Mars and Jupiter. Occasionally an asteroid is knocked out of its orbit and comes close to Earth, bringing with it the chance of a collision.

Large craters on the Earth's surface provide evidence that these collisions have happened in the past. Some scientists think that an Earth–asteroid collision was responsible for the extinction of the dinosaurs 65 million years ago.

Unusual objects which come closer than about 150 million kilometres or so are called NEOs (Near Earth Objects). NEOs are studied and tracked by physicists throughout the world, including the astrophysics research team at Queen's University, Belfast.

Planets have satellites too. Natural satellites of planets are called moons. All planets, except Mercury and Venus, have at least one moon. Saturn has at least 24 of them!

Humans are increasingly putting satellites into orbit around the Earth. These are called artificial satellites and they have four main purposes:

▶ communications
▶ Earth observation (for espionage and monitoring rainforests, deserts, crops, etc.)
▶ astronomy
▶ weather monitoring.

Gravity

The force which keeps the planets and other heavenly bodies orbiting the Sun is the same as that which keeps the Moon and our artificial satellites in orbit around the Earth. It is the force of gravity.

Gravity is a universal force. It occurs everywhere. It exists between all objects that have mass. The more massive an object is, the larger the force of gravity it exerts.

The force of gravity on any object is called its weight. As the force of gravity varies between planets, your weight would change if you travelled to other planets. Table 24.2 shows that the bigger the planet, the more we would weigh. This effect can be noticed very clearly in any pictures of the Moon-landing astronauts. The Moon is about a quarter of the Earth's size and had a much smaller mass. The Moon's gravity is a sixth of that of the Earth and so astronauts weigh six times less on the Moon.

This allowed them to bounce around on the Moon, even though they had extremely heavy suits and packs.

Table 24.2 Weights on different planets

Planet	Approximate weight of 1 kg on that planet/in N
Mercury	4
Venus	9
Earth	10
Mars	4
Jupiter	26
Saturn	12
Uranus	9
Neptune	12

An orbiting astronaut feels weightless in space because there is no external contact force pushing or pulling upon their body, not because there is no gravity! The astronaut is in a state of free fall.

You should remember that on Earth, the weight of a 1 kg mass is approximately 10 N. This tells us the size of the gravitational field on Earth, g, is 10 N/kg. We can therefore write:

weight = mass × gravitational field strength

$$W = m \times g$$

▲ **Figure 24.5** An astronaut experiencing weightlessness in space

Example

1 Find the weight of an object of mass 85 kg a) on Earth and b) on Mars.

Answer

a) $W = m \times g$

$= 85 \times 10$

$= 850\,N$

b) $W = m \times g$

$= 85 \times 4$

$= 340\,N$ (the value for g on Mars came from Table 24.2)

2 An object weighs 1080 N on Venus. On Venus, $g = 9\,N/kg$. Find its mass and weight on Earth.

Answer

On Venus:

$W = m \times g$

$1080 = m \times 9$

$m = \dfrac{1080}{9}$

$= 120\,kg$

On Earth:

$W = m \times g$

$= 120 \times 10$

$= 1200\,N$

Test yourself

1 Name the eight planets in our Solar System in order, from the Sun outwards.
2 Identify the four rocky planets and the four gas giants.
3 State two major differences between comets and asteroids.

Show you can

1 Explain what a satellite is, and describe the difference between a natural and an artificial satellite.
2 State three uses of artificial satellites in Earth's orbit.
3 Suggest why physicists might want to track asteroids that come close to the Earth.

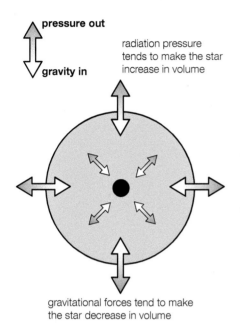

pressure out
gravity in

radiation pressure tends to make the star increase in volume

gravitational forces tend to make the star decrease in volume

▲ Figure 24.6 A main sequence star, like our Sun, is stable because the forces on it are balanced

Stars

Stars are formed in the cold clouds of hydrogen gas and dust known as stellar nebulae. Gravity causes these gas particles to come together. The gravitational force is greater than the outward pressure due to the particles' kinetic energy, and it brings about the gravitational collapse of the cloud. During this collapse, the material at the centre of the cloud heats up as the gravitational potential energy changes into thermal energy. The hot core at the centre of the cloud is called a protostar.

As the protostar accumulates more and more gas and dust, its density and temperature continue to rise, increasing the outward pressure within the protostar. A point is reached where this outward force is balanced by the gravitational force and the protostar becomes luminous because of its extremely high temperature.

If the mass of the protostar is greater than about 8% of our Sun's mass, the temperature will exceed the minimum required for nuclear fusion to begin. There is equilibrium between the inward gravitational force and the outward force from the radiation pressure due to fusion. The star is now in the main phase of its life, so it is called a main sequence star (Figure 24.6).

The temperature at the centre of our Sun is about 15 000 000 °C. This enormous temperature is needed to bring about nuclear fusion. But what is nuclear fusion and why is this huge temperature needed?

Nuclear fusion

In our Sun, nuclear fusion is the combination of two light hydrogen nuclei to produce a single heavier helium nucleus with the release of a vast quantity of energy.

hydrogen nucleus + hydrogen nucleus → helium nucleus + energy

Nuclei are all positively charged particles. They repel each other by electrostatic repulsion. If hydrogen nuclei are to fuse together rather than repel, they must collide at a tremendous speed. This requires the temperature to be very, very high.

Nuclear fusion is the source of energy in every star.

Galaxies

Galaxies are huge collections of star systems. Our own galaxy, the Milky Way, contains over 100 billion stars! These stars are very far apart. The nearest star to us is around 40 000 000 000 000 km away! The nearest galaxy to us, Andromeda, is much, much further away, around 20 000 000 000 000 000 000 km!

The light year

The kilometre is such a small unit that astrophysicists prefer to use the light year.

A light year is the distance that light travels in one year. But how far is that? Light can travel at 300 000 km/s and there are $60 \times 60 \times 24 \times 365 = 3.1536 \times 10^7$ seconds in one year.

So, 1 light year = speed × time

$$= 3 \times 10^5 \text{ km/s} \times (60 \times 60 \times 24 \times 365)\text{ s}$$

$$= 9.46 \times 10^{12} \text{ km}$$

$$= \text{over } 630\,000 \text{ times the distance from the Earth to the Sun}$$

This means that even if we could travel at the speed of light, it would take us 4.2 years to reach the nearest star (other than the Sun). The most distant galaxies are thought to be about 14 billion light years away from us!

Red shift

Think about what we hear when a police car passes with its siren sounding. As the car approaches, the sound appears to have a higher pitch (or lower wavelength) than we would expect. As soon as the car passes, its pitch falls. This is called the Doppler effect (Figure 24.7).

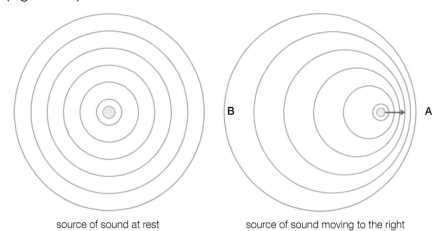

source of sound at rest   source of sound moving to the right

▲ **Figure 24.7** The Doppler effect in sound

The sound source on the right is moving to the right. Observer A hears a sound of high pitch (low wavelength) because the waves are being bunched up together. Observer B, to the left of the source, hears a sound of low pitch (high wavelength) because the waves are being spread out.

A similar effect occurs with light. If the light that we observe from a moving source has a shorter wavelength than expected, it is because the source is moving towards us – we say the light is 'blue shifted'. But if the light we observe has a longer wavelength than expected, it is because the source is moving away from us – we say the light is 'red shifted'.

When we look at the light from distant galaxies, we find that it is shifted to the red (or longer wavelength) end of the visible spectrum. This red shift is due to the Doppler effect.

If there is a red shift in the light from another galaxy, this tells us that the galaxy is moving away from us. The fact that we always get red shift from the distant galaxies tells us that the galaxies are all moving away from us.

The most distant galaxies show the greatest red shift. This tells us that the further away the galaxy is from our galaxy, the faster it is moving away from us.

Big Bang theory

The Big Bang theory is one theory about how the universe began. The argument begins by suggesting that the reason all the galaxies are currently moving away from each other is that they all originated from a common point, called a singularity. About 14 billion years ago, the universe came into existence suddenly with an enormous explosion, which we call the Big Bang. It immediately went into a short period of unimaginable growth, known as inflation.

As the universe expanded, it cooled down and became less dense.

Cooling allowed subatomic particles such as neutrons, protons and electrons to form.

As the cooling continued, protons and neutrons combined to form simple nuclei.

Eventually, around 380 000 years after the Big Bang, and after further expansion and cooling, the first stars came into existence.

There is a further piece of evidence supporting this. In 1964, two American astrophysicists, Penzias and Wilson, detected microwaves of wavelength 7.35 cm that did not come from the Earth, the Sun or our closest stars. The microwaves were evenly spread over the sky, and were present day and night. Penzias and Wilson concluded that these microwaves were coming from outside our own galaxy. These waves come from the cosmos.

Today, we call that radiation cosmic microwave background radiation (CMBR). It represents the signature or 'afterglow' of the Big Bang that occurred 14 billion years ago. Currently, the only model that can give an explanation for CMBR is the Big Bang theory.

Other theories

Today, the scientific community generally accepts Big Bang theory. However, there are other theories. One of these is the **Steady State theory**. It says that the universe never had a beginning because it was always there and always will be there. Physicists are uncomfortable with this theory because it is contrary to the law of conservation of energy. Steady State theory requires us to believe that matter and energy are being created all the time.

Other theories, like the Eternal Inflation theory and the Oscillating Universe theory are variations of the Big Bang Theory.

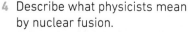

Test yourself

4 Describe what physicists mean by nuclear fusion.
5 State where in the universe fusion occurs naturally.
6 Explain what is meant by red shift.
7 State what is meant by a light year.

Show you can

1 Describe the main features of Big Bang theory.
2 State the main difference between Big Bang and Steady State theories.

1 **a)** What evidence is there that the Earth has been struck by asteroids in the past? *(1 mark)*
 b) Suggest why an asteroid strike today would be so harmful for life on Earth. *(1 mark)*

2 **a)** Describe the main differences between:
  i) the four inner planets and the four outer planets *(2 marks)*
  ii) comets and asteroids *(2 marks)*
  iii) natural and artificial satellites. *(2 marks)*
 b) What name is given to the natural satellite of the Earth? *(1 mark)*
 c) State three uses of artificial satellites which orbit the Earth. *(3 marks)*

3 On the planet Mercury, *g* is approximately 4N/kg. On Earth, *g* is approximately 10N/kg. An object weighs 360N on Mercury. Calculate:
 a) the mass of the object on Mercury *(3 marks)*
 b) the mass of the object on Earth *(1 mark)*
 c) the weight of the object on Earth *(2 marks)*

H 4 **a) i)** What is a light year? *(2 marks)*
  ii) Why is this unit preferred by astrophysicists to the kilometre? *(1 mark)*
 b) The Andromeda Galaxy is approximately 2.5 million light years away from Earth. The speed of light is 3×10^5 km/s. How far away is Andromeda in kilometres? *(4 marks)*

5 Evidence for the expansion of the universe comes from red shift measurements. Explain what red shift means and how it supports the idea that the universe is expanding. *(4 marks)*

6 The most widely accepted model for the formation of the universe is the Big Bang theory. Table 24.3 lists some of the events relating to the formation of the universe, but they are not in the correct sequence. Copy and complete Table 24.3 and order the events by writing a number in the box beside each one. *(3 marks)*

Table 24.3

Event	Order
neutrons and protons are formed.	
rapid expansion and cooling occur.	
further expansion and cooling occur, allowing hydrogen atoms to form.	
more expansion and cooling occur, allowing hydrogen nuclei to form.	

7 According to current estimates, the universe began approximately 14 billion years ago. Explain why physicists believe that the distance to the edge of the visible universe is 14 billion light years. *(2 marks)*

8 Figure 24.8 shows a cloud of gas and dust known as a nebula. The bright spots are stars.

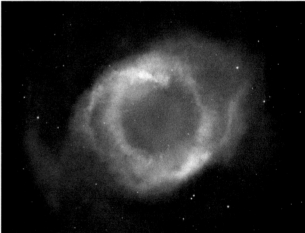

Figure 24.8

 a) What force causes the gas to form stars? *(1 mark)*
 b) What gas is the main constituent of a star such as our Sun? *(1 mark)*
 c) i) Name the process that supplies the energy in stars. *(1 mark)*
  ii) Apart from producing energy in stars, what else is produced by this process? *(1 mark)*

Glossary

UNIT 1

Abiotic factor A non-living factor that can be used to monitor the environment

Active immunity A type of immunity in which the body produces antibodies to combat harmful microorganisms

Allele One of two possible versions of the same gene

Amniocentesis A process by which foetal cells are obtained from the amniotic fluid and then examined for the presence of genetic abnormalities in the foetus

Amnion A lining produced during pregnancy that contains the amniotic fluid

Amniotic fluid The fluid within the amnion that cushions the foetus

Antibiotic A chemical produced by fungi that kills bacteria

Antibiotic resistance An antibiotic-resistant bacterium that cannot be killed by (at least one) type of antibiotic

Antibody A structure produced by lymphocytes that has a complementary shape to (and can attach to) the antigens on a particular microorganism

Antigen A distinctive marker on a microorganism that can lead to the production of antibodies

Association neurone A neurone that connects sensory and motor neurones

Auxin The plant hormone involved in the phototropic response

Benedict's test The test for (reducing) sugar

Biodiversity The range of species in a habitat / ecosystem

Biotic factor A living factor that can be used to monitor the environment

Biuret The test for protein

Bronchitis The narrowing of the airways in the lungs. Usually caused by smoking tobacco

Cancer Uncontrolled cell division

Brown-field site A site that has previously had housing, or other urban structures, on it

Cardiac output The volume of blood pumped by the heart in a minute

Cell membrane The selectively permeable boundary to plant and animal cells

Cell The basic building block of all living organisms

Cell wall A rigid structure immediately outside the cell membrane in plants that provides support. Plant cell walls are formed of cellulose

Central nervous system (CNS) The brain and the spinal cord. The co-ordinators in nervous action

Cervix The opening of the uterus

Chloroplast A structure in a plant that contains chlorophyll. Photosynthesis in plants takes place in chloroplasts

Chromosome A structure found in the nucleus of cells. Chromosomes contain DNA and are sub-divided into genes

Circulatory system The body system that includes the heart and blood vessels

Communicable disease A disease that can be passed from one organism (person) to another

Complication A medical effect that can result from having a condition (such as diabetes) for many years

Condom A barrier contraceptive method

Continuous variation Variation in which there is a gradual change in a characteristic across a population

Contraception A method used to try to avoid pregnancy

Contraceptive pill A contraceptive pill contains hormones and prevents pregnancy by affecting hormone levels thereby preventing eggs being released

Cystic fibrosis A genetic condition in humans caused by having two recessive alleles of a particular gene

Cytoplasm The part of a cell between the cell membrane and the nucleus. Chemical reactions in the cell take place here

Diabetes A medical condition in which the blood glucose control mechanism fails

Diploid The normal chromosome number

Discontinuous variation A type of variation in which all the individuals can be put in a small number of groups with no overlap

DNA The chemical that forms genes and chromosomes

Dominant In the heterozygous condition, the dominant allele will override the recessive allele

Double helix The structure of DNA

Down's Syndrome A genetic condition in humans caused by having one extra (47) chromosome

Effector Muscle (or gland) that can produce a response when stimulated

Egg (ovum) A female gamete

Emphysema Damage to the gas exchange surfaces in the lungs. Usually caused by smoking tobacco

Endothermic A process which uses energy from the environment

Ethanol A type of alcohol, used in the test for fat

Evolution The change in a species over time. Evolution also leads to the formation of new species

Exothermic A reaction that releases energy to the surroundings

Extinction A species is extinct if there are no living members of that species left

Female sterilisation A contraceptive method in which the oviducts are cut

Fertilisation The fusion (joining) of a sperm and egg to form a zygote

Fossil The remains of a living organism that has been preserved (usually in rocks) for millions of years

Gamete Sex cell

Gamete Sex cell that contains only one chromosome from each pair

Gene A section of chromosome made up of a short length of DNA that operates as a functional unit to control a characteristic

Genetic condition A condition caused by problems in genes or chromosomes

 Genetic engineering The deliberate modification of the genome (DNA) in an organism

Genetic screening A process used to test people for the presence of particular harmful alleles or other genetic abnormalities

Genome The entire genetic material (all the DNA) in an organism

 Genotype The genetic make-up of an organism represented by symbols (letters), e.g. tt

Glycogen A storage compound found in the liver consisting of glucose sub-units

 Haploid Half the normal number of chromosomes

Heterozygous The alleles of a particular gene are different (one dominant and one recessive allele)

Homozygous The alleles of a particular gene are the same

Hormone A chemical messenger produced by a gland that travels in the blood to a target organ where it acts

In vitro **testing** Testing of medicines and drugs in the laboratory

Insulin The hormone that lowers glucose levels in the blood

Iodine The test for starch

Leukaemia A type of cancer in which some types of blood cells increase out of control

Lymphocyte A type of white blood cell that produces antibodies

Memory lymphocyte A special type of lymphocyte that can remain in the body for many years

Menstrual cycle The monthly cycle in females of reproductive age that prepares the body for pregnancy

Motor neurone A neurone that carries impulses from the CNS to effectors

Multicellular organism An organism that is formed of many cells

Mutation Random change in the number or structure of chromosomes or genes

Natural selection The process in which the better adapted individuals survive (at the expense of the less well adapted individuals). These individuals are more likely to reproduce and pass their genes on to offspring

Nerve cell (neurone) A specialised cell found in the nervous system. Nerve cells conduct nerve impulses

Nerve impulse An electrical signal that travels through nerve cells (neurones)

Nucleus The control centre of the cell

Oestrogen The female hormone that both causes the repair and build-up of the uterus lining following menstruation and stimulates ovulation

Organ system Organs are organised into organ systems

Organ A structure in the body formed of different tissues to carry out a particular function

Ovary A female organ that produces eggs (ova)

Oviduct Carries eggs (ova) from the ovary to the uterus

Passive immunity A type of immunity produced by injecting antibodies

Pedigree diagram A diagram that shows how a particular condition is inherited through the different generations in a family

Penicillin The first antibiotic developed

Penis Organ that introduces sperm into the vagina

Phagocyte A type of white blood cell that destroys microorganisms by engulfing them and breaking them down (phagocytosis)

Phenotype The outward appearance of an individual, e.g. tall

Photosynthesis A process in plants that uses light energy to produce glucose and starch

Phototropism A plant response in which plant stems bend in the direction of a light source

Placenta The structure that links the uterus wall to the foetus via the umbilical cord. It is here that exchange of materials takes place between the mother and the foetus

Primary consumer An animal that feeds on producers

Producer A green plant that produces food using photosynthesis. An organism at the start of a food chain

Progesterone The female hormone that maintains the build-up of the uterus lining and prepares the uterus for pregnancy

Prostate gland Adds fluid to feed the sperm

Punnett square A grid (table) used to work out the offspring in a genetic cross

Receptor A part of the body that upon stimulation can cause an impulse to be sent to the CNS

Recessive The allele that will only show a characteristic if both alleles are present (and there is no dominant allele present)

Reflex action An automatic response to a stimulus – it does not involve 'thinking time'

Reflex arc The pathway of neurones in a reflex

Respiration A process in living organisms that releases energy

Scrotum Sac that holds and protects the testes

Secondary consumer An animal that feeds on primary consumers

Sensory neurone A neurone that carries nerve impulses from a receptor to the CNS

Side effect An unwanted or unplanned effect of a drug on a person

Sperm A male gamete

Sperm tube Tube that carries sperm from the testes to the penis

Stem cell An undifferentiated cell found in animals and plants that can divide to form other cells (normally) of the same type

'Superbug' A type of bacterium that is resistant to a number of antibiotics

Symptom A sign that shows that something is medically wrong

 Synapse A small junction (gap) between neurones

Testis Produces sperm

 Tissue A group of cells with similar structures and functions

Urethra Tube through which the sperm leaves the penis

Uterus Place where the foetus will develop if pregnancy occurs

Vacuole A large fluid-filled structure in plant cells that contains sap

Vagina Opening of the female reproductive system into which sperm is deposited during sexual intercourse

Vasectomy (Male sterilisation) A contraceptive method in which the sperm tubes are cut

Voluntary action A response that is deliberate and involves the brain making the decision to carry out the response (i.e. it involves 'thinking time')

Zygote The first cell of a new individual following fertilisation

UNIT 2

Acid A solution with a pH less than seven

Acid Indigestion The effect caused by excess acid in the stomach

Activation energy The minimum energy required for a successful collision

Addition polymerisation A reaction in which small molecules are joined or added together to make a long chain molecule and nothing else is produced

Alkali A solution with a pH more than seven

Alkali metal An element in Group 1 of the Periodic Table

Alkaline earth metal An element in Group 2 of the Periodic Table

Alkane A molecule belonging to the homologous series of saturated hydrocarbons with the general formula $C_nH_{2n}+2$

Alkene A molecule belonging to the homologous series of unsaturated hydrocarbons with the general formula C_nH_{2n}

Alternative light source A non-visible light source (e.g. UV light) that can be used to help visualise a fingerprint

Anion A negative ion

Anode The positive electrode in electrolysis

Antacids A medicine used to treat acid indigestion

Atom The smallest part of an element that can exist

Atomic number The number of protons in an atom

Base An insoluble alkali, with a pH more than seven

Biodegradable A substance that can be broken down or decomposed by microbes

Boiling The process by which a liquid turns into a gas

Boiling point The temperature at which a liquid turns into a gas

Catalyst A substance that speeds up a reaction without being used up during the reaction

Cathode The negative electrode in electrolysis

Cation A positive ion

Chemical developer A dye that can be used to help visualise a fingerprint

Chromatogram The visualisation of results of chromatography

Chromatography A separation technique used to separate substances with different solubilities, e.g. different dyes in ink

Collision theory The theory that in order for a reaction to occur particles must collide with sufficient energy

Combustion The reaction of a fuel with oxygen, commonly called burning

Complete combustion The reaction of a fuel with a plentiful supply of oxygen

Compound Atoms of two or more different elements chemically joined together

Condenser (Liebig) The apparatus used to turn a gas into a liquid by cooling

Condensing The process by which a gas turns into a liquid

Covalent bond A shared pair of electrons that holds two atoms together

Covalent compound Two or more atoms joined together by sharing electrons (a covalent bond)

Crude oil A mixture of hydrocarbons formed from the remains of plants and animals from millions of years ago

Crystallisation The process used to separate a dissolved solid from a solvent by heating

Diatomic A molecule containing two atoms

Distillate The part of a mixture separated and collected by the process of distillation

Electrode The piece of apparatus, usually graphite rods, used to conduct electricity into the electrolyte

Electrolysis The process of using electricity to decompose an ionic compound

Electrolyte A liquid that conducts electricity

Electron The negatively charged particle found in the electron shells of an atom

Electron shell The region around the nucleus of an atom in which electrons are found

Electronic configuration The arrangement of electrons in shells around the nucleus

Element A substance that contains only one type of atom

Endothermic A reaction in which energy is taken in

Exothermic A reaction that gives out energy

Filtrate The liquid that passes through the filter paper and filter funnel during filtration

Filtration The process used to separate an insoluble solid from a liquid

Fingerprint The unique pattern found on a person's finger tips

Finite A defined amount of a resource that cannot be replaced once used up

Flame test A test used to identify some metal ions that burn with a specific flame colour

Formula The formula states the ratio of elements found in a compound

Fraction A mixture of molecules with similar boiling points

Fractional distillation A method used to separate miscible liquids with different boiling points

Freezing The process by which a liquid turns into a solid

Fullerene A family of carbon molecules each with carbon atoms linked in rings to form a hollow sphere or tube

General formula The formula of an homologous series that shows the relationship between the number of carbon atoms in the molecule and the other elements

Global warming An increase in the temperature of the Earth's surface

Greenhouse effect The heating effect caused by the layer of greenhouse gases around the Earth

Halogen An element in Group 7 of the Periodic Table

Hazard symbol A symbol used on a chemical to warn of danger

Homologous series A family of compounds with the same general formula and similar chemical properties

Hydrocarbon A substance that contains the elements carbon and hydrogen only

Indicator A chemical that changes colour when added to an acid or alkali

Inert Unreactive

Insoluble A substance that will not dissolve in a solvent

Ionic compound A compound formed from a metal and a non-metal, in which a metal atom transfers an electron(s) to a non-metal atom

Leading edge The top of a spot on a chromatogram

Limewater The chemical used to test for the presence of carbon dioxide

Litmus An indicator made from lichens, which can change between only two colours

Lone pair A pair of electrons not involved in bonding

Mass number The total number of protons and neutrons in the nucleus of an atom

Melting The process by which a solid turns into a liquid

Melting point The temperature at which a solid turns into a liquid

Miscible Two liquids that can mix together

Mixture Two or more substances together that are not chemically joined

Mobile phase The liquid that travels up the stationary phase during chromatography

Monomer A small molecule that will join together in long chains to form a polymer, it usually contains a double bond

Nanomaterial A material made from nanoparticles

Nanoparticles Structures that are between 1 and 100 nm in size, typically containing a few hundred atoms

Natural material A material from a living source, e.g. a plant or an animal

Neutralisation A reaction between an acid and a base

Neutron The neutral particle found in the nucleus of an atom

Noble gas An element in Group 0 of the Periodic Table

Non-biodegradable A substance that cannot be broken down or decomposed by microbes

Nucleus The central part of an atom containing the protons and neutrons

Organic chemistry The study of carbon-containing compounds

Periodic Table The arrangement of all the known elements according to atomic number

pH scale The scale of 0–14 used to measure the strengths of acids and bases

Photochromic material A material that changes colour in response to a change in light conditions

Physical reaction A reaction that is not permanent, it can be reversed easily

Polymer A long chain molecule made by joining lots of small molecules (monomers) together

Polymerisation The process in which a long chain molecule is made by joining lots of small molecules (monomers) together

Property An aspect that describes a substance, for example hard, high melting point, durable

Proton The positive particle found in the nucleus of an atom

Pure A single element or compound that is not mixed with any other substance

Rate of reaction The measure of the amount of reactant used or product formed during a reaction per unit time

Raw material The starting material for a manufacturing process

Reactivity series The order of metals according to how reactive they are with substances such as water or acid

Residue The solid left in the filter paper after filtration or in the evaporation dish during evaporation

R_f value The value given to the ratio of how far a substance travels during chromatography compared to how far the solvent travels

Salt The substance formed during an acid reaction

Simple distillation The process of evaporation and condensation used to separate a mixture

Smart material A material that changes properties in response to a change in the surroundings

Soluble A substance that will dissolve in a solvent

Solute The solid that dissolves in a solvent

Solution The mixture of soluble solid (solute) and liquid (solvent)

Solvent The liquid that dissolves the solute (soluble solid)

Solvent front The distance the solvent travels through the stationary phase during chromatography

State The physical state of a substance, whether it is solid, liquid or gas

Stationary phase In chromatography, the phase that allows the solvent to travel though it. It is used to separate the components in the mixture

Synthetic material A man-made or manufactured product

Thermochromic material A material that changes colour in response to a change in temperature

Trace evidence Evidence that is left at a crime scene, including fibres or hairs etc

Transition metals The block of elements between Group 2 and Group 3 of the Periodic Table

Universal indicator An indicator that can change to fourteen different colours when added to different strengths of acids or alkalis

UNIT 3

Absorber A material which can take in radiant heat

Acceleration The rate at which the speed (or velocity) of a vehicle is changing

Airbags Bags with are inflated by a controlled explosion in the event of an accident. Air bags hold the passenger in their seat to prevent serious injury

Alpha particle A particle emitted in radioactive decay and consisting of two protons and two neutrons

Amplitude The maximum displacement of a particle in a wave from its undisturbed position

Asteroid A very large rock found in space. In our Solar System, many asteroids are found between Mars and Jupiter

Background radiation The radiation all around us coming mainly from outer space and some types of rock below the ground

Balanced forces The forces on an object are balanced if the resultant force is zero

Becquerel A unit or radioactivity corresponding to disintegrating nucleus per second

Beta particle A fast electron emitted in radioactive decay from an unstable nucleus

Big Bang theory The theory that the universe began around 14 billion years ago following a huge explosion

Biofuel A fuel, often used as a substitute for diesel oil, from seeds of certain plants (e.g. rapeseed and castor oil seeds)

Braking distance The distance the vehicle travels after the brakes are applied, until it comes to a complete stop

Cell polarity The existence of a positive and negative terminal in a battery

Coil of wire A length of wire wrapped in concentric rings or spirals

Comet A satellite of the Sun consisting of rock and ice

Compressions Places where the particles in a longitudinal wave are packed most tightly together

Conductor A material which allows electricity (or heat) to pass through it easily. Most conductors are metals

Convection Transfer of heat energy by the movement of molecules of a liquid or a gas

Conventional current The imagined flow of electrical charge from the positive terminal of a battery towards the negative terminal through a circuit

Cosmic microwave background radiation (CMBR) The Background Radiation is the radiation left over from the Big Bang

Crumple zone Part of a vehicle that collapses easily and slowly (crumpling), spreading a collision over a longer time and so reducing the force on the passengers

Distribution Passing electrical energy from electricity pylons to homes and factories where it is used

Doppler effect The change in the observed wavelength of light (or sound) due to the movement of the source

Double insulation A safety system which encloses all conducting parts of an electrical appliance in a plastic box, so that the user can never touch a live component and get an electric shock

Dynamo A device which converts the kinetic energy of a moving magnet into electricity

Earth wire A safety wire which conducts dangerous electrical current away from the metal frame of an electric appliance before it can harm the user

Echo Reflection of waves, usually sound waves

Efficiency The fraction of the total input energy into a machine that is converted to useful output energy

Electric current The flow of charged particles in a circuit

Electromagnetic induction Producing an electrical voltage by changing the magnetic field near a conductor

Electromagnetic waves A family of transverse waves which can all travel trough a vacuum at enormously high speed

Elliptical Not circular, but more like the shape of a rugby ball

Emitter A material which gives out radiant heat

Flash-bang method A method to measure the speed of sound which involves seeing a flash of light and hearing a loud sound

Fossil fuel A fuel, such as coal, oil or natural gas, produced over millions of years by the action of high temperature and pressure on the dead remains of plants and animals

Free electrons Negatively charged particles which are not attached to atoms and are responsible for the conduction of electricity (and heat) in metals

Frequency The number of wavelengths which pass a fixed point in a second

Friction The force which opposes motion

Fuel cells Cells which convert chemical energy in gases directly into electrical energy without the use of a turbine or generator

Fuse A safety device consisting of a fine wire which melts if too much current flows through it, thus breaking an electrical circuit

Galaxy A collection of stars held together by gravity. Our galaxy is called the Milky Way and contains at least 100 thousand million stars. It is thought that there are at least 2 million million galaxies in the Universe

Gamma ray A high energy electromagnetic wave emitted in radioactive decay from an unstable nucleus

Gas giants Planets whose surface is a very dense gas like Jupiter, Saturn, Uranus and Neptune. These planets are much bigger than the rocky planets

Gravitational collapse During star formation a nebula gets smaller and smaller in volume as its particles are pulled together by gravity. This is gravitational collapse

Gravitational field strength A measure of how strong the force of gravity is. On Earth it is about 10 N/kg

Gravitational potential energy The energy possessed by an object because of its height above the ground

Half-life The half-life of a radioactive material is the time taken for its activity to fall to half of its original value

Hydroelectricity Electricity produced by converting the gravitational potential energy stored in the water high above a power station

Insulator A material which does not allow electricity (or heat) to pass through it easily. Gases and most liquids are insulators

Kilowatt-hour (kWh) A commercial unit of electrical energy amounting to 3.6 million joules and equivalent to the energy used by a 1000 watt electric fire in one hour

Kinetic energy The energy possessed by an object because it is moving

Light year The distance travelled by a beam of light in one year. This distance is roughly 9.46 million million kilometres or 5.88 million million miles

Live wire A potentially dangerous wire in a mains electricity supply, usually at a voltage of around 230 V

Longitudinal wave A wave in which the particles vibrate parallel to the direction in which the wave is moving

Main sequence The main part of a star's life. Our Sun is in the main sequence part of its life-cycle

Nebula A huge cloud of gas and dust

Neutral wire A wire used in a mains electricity supply which carries current to the zero voltage terminal

Neutron An uncharged particle (that is, a neutral particle) found in the nucleus of all atoms (except one form of hydrogen)

Non-renewable energy Energy that we can never replace, so we will eventually run out of it. Fossil fuels are non-renewable

Nuclear fusion The joining together of two or more light nuclei (such as hydrogen) to form a heavier nucleus (such as helium) with the release of vast quantities of energy. Nuclear fusion is the process by which stars, like our Sun, get their energy

Nucleus The tiny central part of every atom where most of the atom's mass is to be found. (Do not confuse this with the nucleus of a cell!)

Ohm's Law A mathematical relationship linking voltage, current and resistance, sometimes written $V = IR$

Period The time it takes for one wavelength to pass a fixed point

Power The rate at which energy is transferred or work done. Power is usually expressed in watts or joules per second

Principle of conservation of energy The idea that energy can only change from one form to another, but new energy can never be created and existing energy can never be destroyed

Proton A positively charged particle found in the nucleus of all atoms

Protostar During star formation the collapsed nebula gets very, very hot at its centre. At this stage it is a protostar

Radiant heat Heat given out as radiation from a hot surface

Radiation Transfer of energy by electromagnetic waves from a hot surface

Radioactivity The decay of unstable nuclei by the emission of alpha particles, beta particles or gamma rays

Rarefactions Places where the particles in a longitudinal wave are least tightly packed together

Reaction time The time between a driver seeing or hearing a danger ahead and his muscles responding (e.g. by applying the brakes)

Red shift The light from distant galaxies appears to have a longer wavelength than we would expect. This increase in wavelength is called red shift

Regenerative braking The capture of the kinetic energy lost when a vehicle slows down. This energy is often used to charge a battery

Renewable energy Energy that is produced by nature in less than a human lifetime, so we will never run out of it. Wind energy is renewable

Resistance The opposition by a material to the flow of electrical current. Conductors have a lower resistance than insulators

Resultant force The net force on an object which causes it to accelerate

Rigid passenger cell That part of a vehicle surrounding the passenger that is designed not to crumple in the event of an accident

Rocky planets Planets which have a rocky surface such as Mercury, Venus, Earth and Mars

Satellite An object which orbits another object The Earth is a satellite of the Sun. The Moon is a satellite of the Earth

Solar panel A device which converts the light from the sun directly into electricity or heat energy

Solar system The Sun and the planets, comets, asteroids and everything else which orbits it

Stellar Stellar means having to do with stars

Thinking distance The distance the vehicle travels during the driver's reaction time (while the driver is thinking about what to do)

Transformer A device which converts high voltages to low voltages and vice versa

Transmission Passing electrical energy from the generator in a power station into the wires attached to electricity pylons

Transverse wave A wave in which the particles vibrate perpendicular to the direction in which the wave is moving

Turbine A machine in which wind energy or steam is used to make a large propeller turn round

Ultrasound waves Sound waves which humans cannot hear because their frequency is too high (above 20 000 Hz)

Voltage Voltage causes an electric current to flow

Wavelength The distance between two successive crests or troughs of a transverse wave or the distance between the centres of two compressions of a longitudinal wave

Index